Good Thinking

ALSO BY GUY P. HARRISON

Think

50 Popular Beliefs That People Think Are True

50 Simple Questions for Every Christian

50 Reasons People Give for Believing in a God

Race and Reality

Good Thinking

What You Need to Know to be

SMARTER, SAFER, WEALTHIER, *and* **WISER**

GUY P. HARRISON

 Prometheus Books

59 John Glenn Drive
Amherst, New York 14228

Published 2015 by Prometheus Books

Cover image © Getty Images
Cover design by Nicole Sommer-Lecht

This book is not intended to provide medical advice. Please consult with your physician
regarding health concerns and before implementing changes to your diet and exercise
regimens.

Inquiries should be addressed to
Prometheus Books
59 John Glenn Drive
Amherst, New York 14228
VOICE: 716–691–0133 • FAX: 716–691–0137
WWW.PROMETHEUSBOOKS.COM

19 18 17 16 15   5 4 3 2 1

Library of Congress Cataloging-in-Publication Data

Harrison, Guy P.
    Good thinking : what you need to know to be smarter, safer, wealthier,
and wiser / Guy P. Harrison.
        pages cm
    Includes bibliographical references and index.
    ISBN 978-1-63388-064-1 (paperback) — ISBN 978-1-63388-065-8 (e-book)
    1. Critical thinking. 2. Cognitive psychology. 3. Social sciences—
Philosophy. I. Title.
BF441.H3237 2015
153—dc23
                                                            2015015417

Printed in the United States of America

Dedicated to our ancestors.
Thank you for surviving long enough to get us to here.

CONTENTS

ACKNOWLEDGMENTS

I am deeply grateful to the following people for their valuable contributions to this book: Jade Zora Scibilia, Sheree Harrison, Coni Harrison, Nicole Sommer-Lecht, Catherine Roberts-Abel, Steven L. Mitchell, Peter A. Lukasiewicz, Mark Hall, Fred Gage, Mahzarin Banaji, Jon Simons, Lane Kamp, Benjamin Radford, Kenneth Feder, Nick Wynne, John F. Pfister, Jim Bower, John Byrne, Chanel Prabatah, Cameron Smith, Julien Musolino, and Seth Shostak.

INTRODUCTION

"The fool doth think he is wise, but the wise man knows himself to be a fool."
—William Shakespeare, *As You Like It*

There is good news ahead. This book presents a relatively easy way for individuals to lead safer, more productive, and more rational lives. Good thinking won't necessarily make you more intelligent, but it can make you less prone to making dumb decisions. It probably won't make you rich, but it could prevent you from wasting many thousands of dollars over a lifetime. It can't guarantee you a long and healthy life, but it will make you safer.

The problem of poor reasoning, lack of understanding and appreciation for the methods of science, and ignorance about the structure and function of our brains is the great, invisible crisis of our world. This might seem like hyperbole, but only because bad thinking is so common. It's everywhere, all the time, so we scarcely notice it. Nonsense is our normal, accepted and customary, as much a part of the human landscape as are language and music. We pay a steep price for this tolerance. Irrational thinking slows progress and harms everyone to varying degrees. It doesn't have to be this way. This book lays out a way forward, for both the individual and society.

Critical thinking is the indispensable skill for smart living in modern society, and skepticism is the essential posture for the fully awake twenty-first-century human being. Unfortunately, most people fail to appreciate and utilize either. As a result, they suffer numerous missteps throughout life in the form of wasted time and energy, lost dignity, squandered money, or worse. At the most severe end of the spectrum, countless people bear immense pain and many die prematurely simply because they failed to use their brain well when faced with a critical-thinking challenge. Too many of us believe, trust, buy, invest in, join, or emotionally surrender to the wrong thing at the wrong time.

This book includes an expedition into the brain, specifically your brain. We will consider the long, unpredictable path of evolutionary change that led to our intellectual dominance of the planet. We will discover unexpected ways your brain interprets the world and presents it to you. Veiled zones of the mind await us, where even the most outlandish and ridiculous ideas can become so puffed up and polished that they feel true beyond question. We will confront disturbing subconscious biases against groups of people that may be lurking inside our minds. And we will review current science-based ideas about how best to keep our brains healthy and productive for as long as possible.

Everyone—regardless of talents, academic background, wealth, age, or status—can put the contents of this book to practical use for the rest of their lives. For the sake of our shared future, let's hope increasing numbers of people embrace reason and reality in the coming years. Humankind can do only as well as our collective thoughts and decisions allow. In this age of rapidly expanding technological powers, irrationality becomes correspondingly more dangerous. Good thinking may be the one thing that can save us from ourselves.

Guy
Earth

Chapter 1

THE CASE FOR GOOD THINKING

"For it is not enough to have a good mind; the main thing is to apply it well."
—René Descartes, *Discourse on the Method of Rightly Conducting the Reason, and Seeking Truth in the Sciences*, 1637

*W*ho was this brain? What kind of body was it in? What places had it lived and visited? Was this brain often sad? What made it happiest? Had it been tormented in life, or did it know plenty of joy and freedom? Was it loved? Did it contribute great things to the world? Did it feel the pain of others and try to help? Or did this brain deliver more pain and hate to the world? What were its best ideas? Just how strange and exotic were its wildest dreams?

Those questions and many more bubbled up from my own brain as I stared at the exposed brain on a table before me. I discovered this object of my obsession at the Experimental Biology Conference in San Diego. It was at the booth of a company marketing plastinated body parts for educational purposes. I inspected the details, looked over the brain's blood vessels, the curves and soft valleys of the neocortex. But there was so much more to consider. Is a human brain, I wondered, capable of comprehending the human brain? Is this encounter a variation of the irresistible-force-meets-an-immovable-object scenario?

No mere body part, the brain is like nothing else in the known universe. This is the deepest well of creativity. It is the ultimate organic engine of analysis, perception, and imagination providing us with a borderless expanse of thought. It is, as poet Emily Dickinson wrote, "wider than the sky" because it can imagine, well, just about anything.[1] Regardless of something's size, weight, distance away, or even state of existence or nonexistence, the brain can capture it in a thought. Even something as challenging as infinity can be at least wrestled with and perhaps glimpsed by the brain. Inside it, the unreal becomes real, the impossible possible. Yes, the reach of a single human brain is greater than the bounds of this universe and all others that may be out there as well. But the more I looked at this plasticized brain, the more it seemed small and physically unimpressive. It's certainly not rare, either. More than seven billion of them are operating at any given moment. More than a hundred billion have existed. But there is no denying the magic that is the limitless nature of its output and potential. Somehow, this electrified three-pound blob of water, cells, chemicals, and fat thinks, creates, and dreams in ways nothing else can match. Tens of thousands of miles of blood vessels feed it. Trillions of neural connections give us the ability to walk, run, speak, write, create art, and imagine virtually anything. This one I had found was no bigger than a small cantaloupe, yet it had made and held a lifetime's worth of hopes, fears, dreams, memories, and ideas. A human brain is the greatest asset of all, the most powerful tool ever. And you have one.

Chances are we have not met. But this doesn't mean I don't know a lot about you. I am aware of many of your weaknesses and inclinations because you have one of these human brains. And this tells me more than enough about how you are prone to think and perceive the world around you. Your unique combination of education, personal beliefs, experiences, and personality are of little consequence. You and I belong to the same relatively young and closely related species and rely on the same kind of brain. For this reason, I can say with confidence that you, like me, have at least one foot in fantasyland at all times. It doesn't matter if we happen to speak different languages, eat different kinds of food, listen to different music, hold opposing views on social and political issues, or dress differently. I know that your vision and your memories are not only unreliable but also vulnerable to many forms of manipulation. I know that you feel aware and mostly in control throughout your daily activities. In reality, however, your subconscious mind, I call it the "shadow brain," influences and dictates thousands of your thoughts and actions every day. I know you imagine that you make important decisions based on evidence, logic, and common sense. But the truth is that you, like everyone else, come to many conclusions and initiate many actions based on half-baked hunches served up by your shadow brain. Many of these would embarrass you if they came to light. For example, while you have only begun to shake hands and say hello to someone you are meeting for the first time, your shadow brain is already busy sizing him up and running through a rapid-fire list of questions that will influence your conscious interactions and feelings about this person. *Safe or dangerous? Clean or dirty? Ugly or beautiful? Any potential here for sex or love? Is this person inside or outside my culture/race/religion/worldview? Dumb or smart? Rich or poor?* Factors you would never expect can influence you to like or dislike someone, say yes or no to an investment opportunity or product for sale, and pass a favorable or negative moral judgement on an issue. A smell we cannot even consciously detect, if we happen to be hungry or tired at a given moment, and whether or not we recently washed our hands can tip us one way or another. Regardless of how accurate or fair such rapid-fire hunches may be, they end up in the lap of your conscious mind to deal with. And your conscious mind is more often than not going to accept them without question and begin the automatic process of rationalizing them by selectively noticing all supporting evidence. We can imagine our conscious minds as often slipping into the role of lawyers who have been hired to speak for and defend big corporations, even when the corporations are wrong. We can't stop or cure ourselves of all this subterranean activity—it is normal, not a malfunction or illness—but we can adopt a general awareness about what is going on so that our conscious minds can at least pause and give deliberate thought to the more important decisions in light of all this.

We benefit from understanding as best we can the way our brains work and how we come up with our ideas, conclusions, decisions, and perceptions of reality. Moreover, adopting an attitude of positive skepticism in daily life and striving to be consistent critical thinkers reduces cognitive errors, thereby increasing efficiency and safety. Few things in life as this can offer more protection from harm and more opportunities for personal and professional fulfillment. Sadly, however, relatively few people ever make the effort to learn how brains work, how to keep them healthy and productive for as long as possible, and how best to think through the mental land mines lying around within and without the brain. Failing to know one's own brain is a dangerous ignorance. Bad thinking is our invisible pandemic, the plague popular culture ignores. The invaluable bundle of critical thinking, skepticism, appreciation for science, and a basic understanding of the human brain fits nicely into one simple term: *Good thinking*.

WHAT IS GOOD THINKING?

Good thinking is an umbrella term for understanding the human brain and using it in ways that enable one to make rational decisions, identify deception, and avoid or discard delusions as often as possible. It requires the following:

- An understanding of the evolutionary history of the human brain and how it has left us with a thinking organ that goes about its business in unexpected ways that mislead us about reality.
- Knowledge of the basic structures and functions of the human brain, how vision and memory work, for example. How personal recollections and sensations can seem real even when they are not.
- An appreciation for the profound impact nutrition, life-long learning, and physical activity have on the brain's health, performance, and longevity.
- Awareness of the prominent role of the subconscious mind in daily life, and the understanding that we inherited our brains from ancestors shaped by extremely competitive and dangerous environments that made fast subconscious reaction a priority over slower conscious reflection and imagination.
- An alertness to many of the natural and common mental biases and shortcuts that can undermine rational thought.
- The courage and maturity not only to question everything but also to accept the absence of answers and those answers that may contradict hopes and beliefs that appeal to us.
- Sufficient humility to prevent one from placing absolute trust in sensory perceptions, personal experiences, and even thoughtful conclusions. A willingness to always reconsider, revise, and change one's mind when better evidence demands it.

There is a growing global subculture of people who can be described as *good skeptics*. They are not cynics or pessimistic malcontents. These are people who simply have decided to think before they believe, to question strange or important claims before accepting them. They appreciate the protective value of critical thinking. They ask for evidence and look for logic and reason before accepting a claim. Good skeptics are nothing more than people who strive to be sensible. They feel it's wise to put in a bit of work in order to spot as many lies and delusions as possible. Who in their right mind wouldn't want this? The better one is at separating truth from fiction, reality from delusion, the better chance one has to avoid problems, because our world overflows with bad ideas and deceptive claims. Of course, many things defy easy categorization. No one can tell for sure whether or not something is true or real in every case. What then? A good skeptic understands this and is necessarily humble as a result. There are more questions than answers, more mysteries than solutions. Good skeptics are not so silly to imagine themselves as know-it-alls who have everything figured out. They are honest enough to admit ignorance when appropriate and to live with unanswered questions—even when doing so may be unsatisfying or distasteful. This is a subculture open to anyone, anytime. It's never been easier. There is no entry test, no application to fill out, no membership fees, no secret handshake. All it takes is a moment of clarity, a decision made within the privacy of one's mind to embrace skeptical thinking.

One of the reasons it is easier than ever to be a good skeptic and rational thinker these days is because we know so much more about ourselves, thanks to science. The ways in which we see, hear, process information, remember, and make decisions are no longer as mysterious and misunderstood as they once were. Much remains to be learned, of course, but the fact is hard work over many years by many scientists has revealed a remarkable amount of practical information about our brain. Why not take advantage of it to better your life? Unfortunately, most people don't know or don't bother. As a result, they inevitably become the unwitting victims of their own brain's natural functions again and again. Possibly without meaning to, they cling to nonsense and habitually make bad decisions. And once negative consequences come home to roost, they

blame it on bad luck, their enemies, or perhaps a divine plan. Ignorance about what the physical brain needs to work well and last
long is also the norm. For some, the cost of all this may be merci-
fully minor. For many others, however, it might cost everything.

"It's all about self-awareness," said Kenneth Feder, an author
and professor of archaeology at Central Connecticut State who
promotes science, skepticism and critical thinking.[2] "Most people
merely assume that their brains objectively collect data through the
senses. I think it's humbling, in a good way, when people learn it's
not nearly that simple. Knowledge of and an understanding of pareidolia, cherry-picking, confirmation bias, [and] the ways our brains
fill in missing information are all useful in recognizing our limitations and in recognizing that there is a viable process of increasing
objectivity—science."

When people reflect on poor decision making, bad judgment,
prejudices, and irrational fears, they are likely to view these as
information deficiencies and problems *other* people struggle with.
It is common, probably universal, for people to think that they are
more rational and careful in their thinking than they really are. If
only *other people* knew more facts or were more educated, then they
would make better decisions, be less prejudiced and less prone to
being taken advantage of by charlatans, right? Not necessarily. It's
not just "them." It's all of us. After countless experiments and many
thousands of published studies, we know enough about the brain to
be sure that no one is ever safe from cognitive deceptions and just
plain sloppy thinking. Scientists know this. But what about the rest
of us? At the very least, make sure *you* know.

APPLY SCIENCE IN YOUR LIFE

At its most basic level, science is a process of discovery available
to all of us. The scientific method can be applied to many of the
challenges we face in our everyday lives. Here are the steps:

1. Research, observe, collect information.
2. Develop an idea, hypothesis, or theory that might explain,
   confirm, or cast doubt on the idea, claim, product for sale,
   or challenge in life.

3. Design and execute an experiment or test that might confirm or disprove your hypothesis.
4. Evaluate the results, repeat the same experiment or conduct different experiments if necessary.
5. Share findings with the smartest people you know. Ask them to look for errors in your reasoning.

I'll make one more assumption about you: You think you are pretty smart, maybe not a genius, perhaps, but definitely above average. If I'm right about this, and you know I am, it's crucial for you to understand that being intelligent doesn't make you a good thinker. A respectable formal education won't necessarily do it either. It's not enough. For example, it is only natural for all of us to underestimate how our judgments are shaped and directed by hidden biases. From buying soap to voting in an election, we wrongly assume that we make decisions based on rational thinking—even though we can see that so many others around us do not. Research has shown that this is how people typically think.[3] We simply aren't adept at recognizing how bad we are at critical thinking. But we can improve. The first and possibly most important step toward good thinking is humility. Believing that you are rational and your conscious self is in control of your perceptions and decisions is itself the greatest of all delusions. Let go of that fantasy and admit that you, like every other human on Earth, need to work on your good thinking.

Fortunately everyone is already a skeptic, and we are all critical thinkers. Some people certainly appear to try, but no one embraces *every* crazy claim. No one falls for *every* scam. This means that in all of us there is at least a foundation, a place from which to build. The bad news, of course, is that in many cases some heavy-duty construction is necessary. Becoming a better critical thinker means improving what you already do to some degree, not changing everything about how you think and feel. I have yet to meet someone who told me that believing nonsense or wasting time and money on lies is a great way to live. In virtually every case, people accept bad ideas only because they haven't figured out that they *are* bad ideas. No one sets out to make mistakes. No one yearns to be a sucker. Virtually everyone is in the game; we just need to play it better.

THE DIRTY DOZEN

Know these twelve common mental mistakes and resist them when they threaten to hijack rational decision making. Be on guard when others use them in attempts to convince you to accept unusual claims.

1. **The Emotion Potion.** We are emotional creatures, and this often leads us to make irrational decisions, embrace bad ideas, and act in ways that work against our best interests. Emotions can intoxicate us, make us dumb. Be aware of this vulnerability. If someone tells you the world is going to end in fiery chaos soon, for example, don't let your fear of such an event distract you from rationally analyzing and challenging the claim.

2. **Popularity.** We are social animals. The safety of the herd feels good, and it can be cold and harsh out there all alone in the wilderness. Recognize how we all can be swayed by popular support of an idea no matter how destructive or ridiculous it may be. Never forget that truth and reality are not decided by vote. The majority of people have been wrong about many things many times throughout history. There were times when a flat Earth, phrenology, and bloodletting were respectable and popular ideas—but they were still wrong.

3. **Straw Person.** A common tactic people use to promote weak or worthless claims is to attack an easy-to-beat, diluted, or counterfeit version of the counterargument. Those who say, for example, that Earth is around 4.5 billion years old should not be swayed one bit on this point if a science denier were to tell them that there was a time when geologists didn't understand continental drift and still can't explain everything today about the structure and function of the Earth's core. Of course geologists don't know everything. But this does not refute the strong evidence for a 4.5-billion-year-old Earth.

4. **Loaded Questions.** Sometimes people try to make their point seem more sensible by slipping in an unproven claim or bit of nonsense as filler or padding. Example: "Another reason we know the Lost City of Atlantis is real is because psychics and mediums have communicated with dead

Atlanteans." Listen well and catch weak arguments or bad ideas within the larger claim. Challenge them all.

5. **Wishful Thinking.** Simple but deadly to good thinking. We desire something, so we believe it to be true. This is a powerful human compulsion. Be aware of it and be tough with yourself. Always ask, "Am I accepting this claim because it makes sense and it is supported by sufficient evidence? Or do I just want to believe it so much that I am willing to pretend to know it's valid?"

6. **False Dilemma.** Watch out for people who frame their case as an "either, or" proposition. Sometimes there is a third option, or perhaps many more options. For example, a politician might say that more prisons must be built or there will be more violent criminals on the streets. But what if nonviolent offenders were released early or given lighter sentences, freeing up space for more dangerous criminals to serve longer sentences?

7. **Explaining by Naming.** Giving a name to something is not the same as explaining it. For example, calling an event a "miracle" is not an explanation for what happened. Calling a session with a psychic a "reading" does not explain how information was supposed to have been retrieved by the psychic. Watch for this deceitful form of verbal carpet bombing and simply ask people to explain names and concepts they try to pass off as explanations.

8. **Circular Reasoning.** Always popular in religious circles, this one also gets plenty of mileage in other arenas as well. It happens when people attempt to prove A by pointing to B, which they claim was proved by A. Example: "My special book is true because it was inspired by the gods and I know the gods are real because my special book says so."

9. **Authority Worship.** Try to remember that in many ways we are essentially chimps who wear shoes. Just like them, we are obsessed with social rank and power. This is a huge weak point in our brains, because our natural reaction is to snap to and obey when we view someone as our superior. I'm not suggesting you rebel against everything every authority figure says to you, of course. But do try to think clearly about the validity of words from on high. Don't let a uniform or someone's dominant posture hoodwink you into believing nonsense or buying a junk product.

10. **Special Pleading.** People who promote or believe in things that are unlikely to be true often scramble to change the game when they feel the walls of reality closing in on them. For example, a person who says acupuncture works because "one billion Chinese people can't be wrong" might not like hearing that only about 18 percent of China's population relies on acupuncture[4] and react by arguing that numbers suddenly don't matter.

11. **Burden of Proof.** The person making the extraordinary claim is the one who has the responsibility of backing it up. You and I don't have to prove that mediums can't talk to dead people or that aliens have never visited Earth. It's not even fair to suggest that we should, because in most cases it would be impossible to definitively disprove such things. Instead, it is the believer who must validate her beliefs.

12. **Ad Hominem Attacks.** When one can't get anywhere attacking facts, the next best thing seems to be attacking the person aligned with those facts. This is a weak, immature, despicable tactic—and we all do it. But that doesn't make it right. If you are discussing astrology with a jerk, remind yourself that being a jerk is irrelevant to whether or not astrology's claims are true. Focus on logic and evidence. It's better for everyone in the long run to kill the message rather than the messenger.

"In daily life, critical thinking is our only defense against being fleeced," said Portland State University anthropologist Cameron M. Smith.[5] "Without it we are easily fooled by politicians and snake-oil salesmen to name just a few. In science, a healthy, constructive skepticism is required to sort the good theories from the bad and slowly, steadily make our way to understanding."

In childhood we learn to rely on sight, sound, and touch to make sense of the world. Unfortunately, no one teaches us that our senses can mislead us about what is real and what is not. We also have many biases and ways of thinking that feel right but steer us toward flawed conclusions and poor decisions.

"Do not trust your senses!" warns John Pfister, senior lecturer in the Department of Psychological and Brain Sciences at Dartmouth College.[6] "Evolution has shaped us to be big gap-filling machines and this can get us into trouble over and over again. 'Goodness, I dreamed of Uncle Floyd last night, something must have happened!' 'Holy cow! There I was thinking about Uncle Floyd the other day and he called me! We must be psychic.' 'Did you hear that Uncle Floyd came out of his coma last night? Our prayers must have been answered!' No, just because something feels like it should be true does not make it so. Intuition, while a good starting gun for the race to find the truth, is only that. Empiricism, logic, and the scientific method will carry you over the finish line."

Forget how much general knowledge you have or how many diplomas you may have racked up. It doesn't matter. Good thinking comes down to performance and not credentials. Sound information is nice to have, but knowing how to think matters most. Good thinking is not the same as being smart or educated. It means mental clarity, agility, and dynamism. It's a form of intellectual due diligence concerned with doing the work to determine what is worth knowing and accepting as probably true. What is real? What is true? Will this product deliver as advertised? Is the source credible? What are some appropriate expert sources of information about this that I can consult? Should I just admit I don't know and leave it at that for now? Good thinking requires one to carefully consider conclusions and decisions before, during, and even long after they are made.

"Chances are pretty good that most self-identifying skeptics did not learn their critical thinking skills as children in school," said John Byrne, a Michigan internist, pediatrician, and assistant professor at the Oakland University William Beaumont School of Medicine.[7] He thinks everyone, the younger the better, should learn as much as possible about the human brain.

"I think that it is also likely that most self-identifying skeptics wish that they had some formal education in critical thinking early on. We have learned a lot about how wrong our instincts are concerning how the brain works. Everything we experience is really our brains' interpretation of the world. Our biases are built-in filters that prevent us from seeing some of the world as it really is. Information is processed to make sense of the world with respect to our previously held beliefs even before we are consciously aware of it. This is apparent when two people observe the same event yet have two distinct memories of the event. It is also obvious when we study optical illusions."

Years ago, Byrne performed as a magician at children's parties, where he saw firsthand how eager children were to learn about the many unexpected workings of the human brain.

"Children naturally love illusions," he said. "They get a kick out of being deceived. And they want to learn how and why they were deceived. It would seem that children would be excited to learn about how their brains process information. Of course, this excitement could only be fostered by teachers who are not only well trained in critical thinking, but are passionate about it. By beginning with the knowledge that our brains are not really concerned so much with truth but rather with preserving our worldview, children would learn to look at the world differently."

Byrne continued, "If we could foster that excitement early on, our children might grow up with a healthy sense of skepticism. Critically thinking children grow up to be critically thinking voters, consumers, and caregivers. Most people are naturally compassionate, but compassion can be derailed by flawed notions of how the world works, especially in healthcare. Children raised with compassion combined with healthy skepticism could lead to a world with less nonsense standing in the way of progress toward a happier, healthier society."

One might think that most people would recognize the value of brain awareness and critical-thinking education for everyone, especially the young. But I'm not so sure. Some may view good thinking as a threat to their worldview. In the United States, for example, three in four adults believe in things such as haunted houses, astrology, and spirit channeling.[8] Forty percent of people in Great Britain believe in haunted houses.[9] Nearly a quarter of Canadians think some people have the ability to carry on two-way conversations with the dead.[10] Keep in mind, the United States, the United Kingdom, and Canada are among the highest ranked nations in the world for education, and all three have literacy rates of at least 99 percent. Still, the optimist in me demands we remember that no one believes *everything*. Even most Loch Ness monster believers probably draw the line at leprechauns. Everyone has at least some critical-thinking skills to deploy sometimes. We don't have to necessarily reinvent ourselves as a species, just get better, if that helps make the challenge before us feel any less daunting.

Bad thinking leads people to believe first and ask questions later—if ever. Good thinking enables people to duck and dodge nonsense, both trivial and dangerous. It allows one to weed the garden of the mind and keep it fertile for useful work. The challenge is persistent and inescapable. It exists throughout all societies, and the toll is great, down at the level of the individual and all the way through to the flow of historical progress. If only critical thinking and brain education were on the short list of human needs and values that crossed all borders. Imagine: Clean water, sanitation, nutrition, healthcare, and *good thinking*. What a world we might build for ourselves. Pfister agrees that valuing and pursuing a general education about our brains is a no-brainer:

> It is not like you can avoid having one, or give it to another person to keep an eye on. No, it is yours and yours alone to enjoy, be frustrated with, and marvel at its wonders. That being the case, you should also read the owner's manual. You will learn how to use it wisely—calculate, ponder, imagine; what not to do with it—drugs, directly expose [it] to sun or anything else toxic, for that matter; how to keep it safe by wearing a helmet; what it is not good for—remembering things like a videotape recording, understanding randomness and coincidence, betting on the lottery. No, teaching students Algebra II might be fine for some places, but I would hope that a course in critical thinking might breed a new generation of responsible citizens, eminent scientists, and reasonable policy makers.[11]

Why do so few people know about critical thinking, skepticism, and brain anatomy, given the direct importance of these things to the quality, productivity, and safety of their lives? Considering the disturbingly high rates of irrational belief in contemporary society, it is safe to assume that few parents sit down with their young children for serious talks about the value of good thinking. Always the emphasis at home, just like at school, seems to be on *what* to know rather than *how* to know. It is the rarest of schools that teaches critical thinking to all students. Those who are fortunate enough to ever receive instruction in critical thinking or brain science usually do so only at the university level. And even then, it is probably but a minority of students who are majoring in a directly related field such as psychology or biology. Outside of philosophy majors, most university students probably never take a class that is specifically designed to teach them how to become better lifelong thinkers. Do any high schools or middle schools anywhere require all their students to pass a critical-thinking course in order to graduate?

Basic brain education is also inexcusably missing. The world's schools do not do an adequate job of teaching children about the human brain. Most students spend more than a decade of their lives in classrooms without ever hearing or reading much, if anything, about neurons, glia cells, the hippocampus, and so on. It's likely that if children understood how the brain evolved, how it learns, why food and sleep matter to it, and so on, they might be better students. And this likely would be good for the entire world. Fred Gage, a leading brain researcher at the Salk Institute in La Jolla, California, thinks so. "I believe there would be a positive effect on children if they were taught about basic brain structure and function and exposed to critical-thinking concepts. If they apply the knowledge they gain in a positive way, this would benefit society."[12]

The result of all this educational neglect is that relatively few of the adults who currently head families, businesses, and countries were ever exposed to—much less learned and embraced—ways to use their brains well and to rationally think through important ideas and claims. The harsh reality is that most people are left to discover good thinking independently. There are many excellent, free resources online, but only for those with access to a computer and the Internet. And even if access is available, one must stumble upon the thought that brain science and critical thinking are worth researching. The majority of people never seem to encounter that thought. After all, it's not something cultures typically value and promote. Imagine if we left literacy or math competency to be achieved in this way. *Just leave kids to figure out multiplication on their own; they'll sort it out eventually. Who needs language arts in middle school? Children can work it out.* The consequences would be unacceptable, would they not? So why aren't children taught about their brains and good thinking as a matter of mandatory routine in every school? Shouldn't we all receive a basic education early in life about the one thing that we and civilization depend upon most? Isn't the structure and function of the human brain at least as important as any other standard school subject? Learn to read. Learn to write. Learn to add and subtract. *Learn to think.*

"A better understanding of essential brain anatomy and function could help to clarify students' understanding of the source of their own thoughts and feelings," said anthropologist Smith.[13] "The human brain and mind appear to be the most complex structures produced so far in the history of Earth evolution, and the more people knowledgeable about them only increases the chance that we'll better understand our own minds."

Is there something about good thinking—maybe the asking questions and requesting evidence part—that scares us? Are we afraid of losing something if we dedicate ourselves to reason and reality? Is this our excuse for not making good thinking a rite of passage for all? Professor Feder likes the scene from the 1960 film *Inherit the Wind*, the film that portrayed the famous Scopes Monkey Trial of 1925. Matthew Harrison Brady (the character based on William Jennings Bryan) asks Henry Drummond (Spencer Tracy/Clarence Darrow) if he views anything as sacred. Drummond responds:

> Gentlemen, progress has never been a bargain. You've got to pay for it. Sometimes I think there's a man behind a counter who says, "All right, you can have a telephone; but you'll have to give up privacy. . . . Mister, you may conquer the air; but the birds will lose their wonder." Darwin moved us forward to a hilltop, where we could look back and see the way from which we came. But for this view, this insight, this knowledge, we must abandon our faith in the pleasant poetry of Genesis.[14]

"I admit," said Feder, who is also a *Skeptical Inquirer* consulting editor, "there is a part of me that misses the certainty I had as a kid, that there was a god and a heaven and that I would someday reunite with deceased loved ones, which at that point was just a dog."[15] He continues:

> But, in trade, I get to marvel not just at what the world looks like—the night sky far away from the ambient light near a city; the multicolored spires of Bryce Canyon; tigers; human babies; a beautiful woman—but also to understand on some level the amazing processes involved in their existence. It's funny; I will sometimes get a student who'll say something like: "But wouldn't it be interesting if ancient astronauts landed on Earth and helped the ancient Egyptians build their pyramids?" Well, of course that would be cool. But wouldn't, on some level, the pyramids lose some of their grandeur? I mean, they wouldn't be all that surprising or impressive if beings who were capable of building starships could pile up big stones very accurately. But that fact that "mere" humans could, with just their own intelligence and muscle-power, now that's impressive.

The results of our negligence to teach, encourage, and expect good thinking are clear to see. The world is a swirling, festering ball of deceptions and madness that all of us must wade through each day. For anyone unconvinced that poor thinking skills and weak skepticism are the great unrecognized crisis I claim, let's briefly consider a few statistics. Sadly, what should be fringe beliefs virtually extinct by now are still mainstream in twenty-first-century America and many other societies.

According to a Harris poll, 42 percent of American adults think ghosts exist despite the fact that in all the centuries this claim has been haunting us, no one has ever produced any good evidence for it.[16] A Pew Research Center study found that 33 percent of adult American women say they have had at least one personal encounter with an animated or conscious dead person. Twenty-six percent of men make this same claim.[17] After all this time, don't you think science might have produced some—any—good evidence for ghosts if they were real? I'm not suggesting that the existence of ghosts has been disproved, only that there is nothing to justify belief in them.

By the way, I will praise science more than once in this book for its superior ability to reveal reality and expose mistakes in our thinking, so I should make it clear to readers that its usefulness and value do not suggest perfection. Far from it, science is a castle built upon mistakes. Scientists make errors all the time. The key is that sooner or later they acknowledge stumbles and dead ends and learn from them. "One of the beauties of science is that it has built-in error-correcting machinery," the late astrophysicist and science popularizer Carl Sagan explained in a 1996 *Psychology Today* interview.[18] "Science, unlike many other human endeavors, reserves its highest rewards for those who disprove the contentions of its most revered leaders. Think, for example, of religion. How foreign that scientific point of view is from the religious idea, which is so often to uncritically accept whatever the founder of the religion said. It's not a tragedy that scientists make mistakes, and I certainly have made some in my time."

How old is our planet? How old is humankind? Bad thinking gets in the way of even the most basic facts about our existence, which leads to an intellectually diminished life. Today more than 40 percent of American adults think all life, including the human species, and the Earth itself are less than 10,000 years old.[19] This claim is manifestly absurd, as numerous branches of science have revealed an abundance of converging evidence that points to anatomically modern humans evolving from ancestral species approximately 200,000 years ago and an Earth that is some 4.5 billion years old. These are not beliefs or magical revelations. They are not to be accepted because someone in a position of authority says so. They are *evidence-based conclusions*, subject to change if better evidence emerges. Believing Earth popped into existence less than ten millennia ago is about as far off the mark as believing that the distance from Earth to the Moon is a third of a mile, or that the distance between Los Angeles and New York City is twenty feet. Bad thinking dims our view of the universe and ourselves. It robs us of the human story.

Good thinking means understanding that our desires push us to believe in things well before they deserve it. We must always evaluate and question our thoughts and conclusions in order to prevent what may be reasonable hopes from mutating into unreasonable beliefs. For example, there are good reasons to give serious consideration to the possibility of intelligent extraterrestrial life somewhere in the universe. If there is life beyond the Earth, intelligent or otherwise, I certainly hope it is discovered in my lifetime. A few years ago, I visited the SETI Institute (Search for Extraterrestrial Intelligence) in Mountain View, California, to be interviewed on its radio show, *Big Picture Science*. I felt like a kid on Christmas morning who had been teleported into a candy store. Just seeing the nameplate on SETI founder Frank Drake's office door, for example, brought back memories of my first encounter with the Drake equation when I was a young child and how it inspired me to think deeply about the possibility of alien life. I grew up but never outgrew those thoughts. I *want* intelligent extraterrestrials to be out there. I *want* us to find them. But all my hope and enthusiasm is no match for my good thinking. No matter how good it may feel, I refuse to pretend to know something I do not know, and at this time there is no scientifically verifiable evidence for alien civilizations or visits by their emissaries. Therefore, claiming to know it would not be rational. Nonetheless, 32 percent of Americans do believe in "UFOs."[20] Because such believers are weak skeptics with poor or inconsistently applied critical-thinking skills, they assume that seeing something strange in the sky or hearing a good story about alien visitors is enough to prove the claim. But certainly it is not.

FROM ASTROLOGY TO ZOMBIES:
INTERVIEW WITH THE SKEPTIC

Benjamin Radford is a well-traveled expert in the world of weird beliefs and hollow claims.[21] As a professional skeptic and paranormal investigator for more than fifteen years, he has just about heard it all, from astrology to zombies. Radford has also authored several books and is currently deputy editor of Skeptical Inquirer *magazine.*

Q. After all these years of investigating extraordinary claims and irrational beliefs, why do you think so many normal, smart people fall for nonsense?

A. My educational background in psychology has been enormously helpful to me as a skeptic and science-based investigator. The reason is simple: whatever the topic, whether it's miracles or Bigfoot, psychics or ghosts, the vast majority of the evidence offered for these supposedly mysterious and "unexplained" phenomena is in the form of a personal-experience story. Look at it this way: If there was clear, incontrovertible scientific evidence that these things existed, then they would not be "mysterious" or "unexplained" at all—instead they would be obvious to anyone, regardless of belief or opinion. There would be no question whether or not ghosts or Bigfoot exist, because anyone could walk into a natural history or science museum and see one. They'd be on display, like any other creature, energy, or entity. We'd know what their properties and limits are, what they do under certain conditions, and so on.

What you find is that the vast bulk of the evidence offered for virtually any "mysterious" phenomenon consists of anecdotes—stories or personal experiences. When I interview people, the answer is almost always, "I *know* it exists because I saw it, it happened to me. . . ." Only rarely do believers try to cite any sort of research or scientific evidence.

Skeptics and psychologists, of course, are very aware of the limits and fallibility of eyewitness and personal experience, since our perceptions and recollections are subject to many different sources of bias. This doesn't mean that our personal experiences are worthless, of course—just that they are not necessarily a reliable guide to what really happened. Studies show that the public generally greatly overestimates the validity of personal experience, which is understandable since we rely on personal experience for most of our waking hours!

Q. What are your thoughts on the human brain? Do you feel we are using it incorrectly?

A. Since so much of the evidence—and I use that word in its broadest possible sense—for the paranormal is rooted in the brain, it's important to understand those perceptual, cognitive, analytical, and other processes. It's not just an abstract, theoretical question; I see it in nearly every investigation I've done. Since we are humans using fallible human brains to understand the world, including evidence offered of experiences reported by other humans using their own fallible brains, you can't really get away from it.

The human brain likes certainties, it likes simplistic, binary thinking. It has a strong tendency to categorize the world into A and B, safe or unsafe, yes or no. But the real world is full of conditions, caveats, and nuances. It's very difficult to keep all of the important factors of a given situation in mind at one time, and our brains understandably want to simplify. In my opinion and experience, that mistranslation between the real world and our comprehension of that world and our need to generalize are often at the root of misunderstandings. My feeling about the brain specifically is that it's not so much that people are using their brains incorrectly, it's that most people aren't really taught how to overcome, minimize—or even more importantly *recognize*—these common logical errors and cognitive biases. That's what good thinking is all about.

The best you can do is apply scientific methods to control for those factors and biases as much as possible: You use experimental research-design elements like control groups, randomization, blind and double-blind studies, and that sort of thing. No experiment is perfect, and no result is scientifically definitive, of course, but science doesn't deal with absolute certainties. The scientific methods help us to tease out significance from chance, signal from noise.

Q. Are you optimistic or pessimistic about humankind?

A. I'm a generally optimistic person, and I'm often puzzled by the gloom-and-doom scenarios I hear people discuss as an inevitable path for humanity. Yes, there are serious environmental and other issues that need to be addressed, but the world is not going to hell in a handbasket. We are fallible, and we will of course never be rid of the violence, prejudice, and superstition that plague the human condition. But if we can help encourage critical thinking and evidence-based reasoning, that can certainly help provide a better future for all of us.

When we allow our desires to have their way with us, silly ideas can seem a lot like serious, true knowledge. Here in the early years of the twenty-first century, for example, a large number of people worldwide actually think that the positions of stars and planets at the moment of a person's birth fix their personalities and cast a permanent shadow of influence over their lives. More than a quarter of American adults believe in astrology today.[22] There is no evidence for astrology, no logical arguments, nor credible research to support it. Yet the belief thrives generation after generation. In the United Kingdom, 24 percent are under the spell of star power.[23] Indian astrophysicist Jayant V. Narlikar describes astrology belief in India as "almost universal."[24] I can vouch for its popularity there, having visited India and poked around in many of its markets and bookstores. Worldwide, I estimate that at least four to five billion people, easily more than half of the population, think astrology is true to some degree. Yes, even as our species successfully lands robotic probes—nonbiological extensions of our biological brains—on distant worlds and inches ever closer to understanding the secrets of nature around us and within us, billions of people still look up at space and see a vast zone of magic. And it is all supposed to work by some mysterious force that neither scientists nor astrologers can detect, measure, or explain. My first question to those who profit from or enthusiastically promote this stuff is simply this: What is the theory of astrology? I just ask them to explain how it works. They can't, of course, because it doesn't work and it's based on nothing more than the wild ideas of ancient people who looked up at the night sky and let loose their imagination. Meanwhile, professional astrologers today prey on weak skeptics who do not understand how the human brain is naturally attracted to simple explanations of complexity and how it can make ideas that feel good also seem sensible. These outer-space soothsayers extract tens of millions of dollars annually in exchange for "carefully calculated" predictions and insights—even though their work is missing an explanatory theory and is based on outdated and inaccurate positions of celestial bodies.[25]

How do we explain the popularity of something like astrology when it is so easy to punch holes in its claims and expose it as worthless pseudoscience?[26] It is not as simple as blaming a lack of formal education, and we certainly can't pretend that rampant mental impairment is driving its popularity. Runaway belief in astrology and other baseless claims is the inevitable outcome when critical thinking is rare and inconsistent. Throughout history and today, many astrology believers have been brilliant and highly educated people. It has reeled in all kinds, from paupers to presidents.[27] Smart people are seduced by dumb pseudoscience every moment of every day. But it doesn't have to be like this.

The only solution to the problem of bad thinking is good thinking and all it entails. Traditional education cannot inoculate us from irrational thinking, nor can wealth, status, or achievement. It may help some, but not nearly enough. Let's consider the case of Britain's Member of Parliament David Tredinnick. He is a graduate of Eton, one of the United Kingdom's most selective and expensive boarding schools, as well as St. John's College, Oxford. St. John's College is part of venerable Oxford University and describes itself as a school that "fosters critical thinking, creativity and excellence both inside and outside the classroom."[28] He also has an MBA from the University of Cape Town, South Africa. Nonetheless, Tredinnick somehow ended being a proud and outspoken advocate for astrology. As a Member of Parliament with a seat on the British government's *science and technology* committee, Tredinnick once told his Parliament colleagues that he is "absolutely convinced that those who look at the map of the sky for the day that they were born and receive some professional guidance will find out a lot about themselves and it will make their lives easier."[29] Apparently he doesn't know or doesn't care that astrology's claims have never been confirmed by the scientific process or that the world's astronomers and astrophysicists—the people who know more about planets and stars than anyone—reject astrology as a patently false claim based on absurdities. No doubt Tredinnick is smart, successful, and well educated. But his unwavering belief in astrology suggests that he has not yet embraced good thinking.

Everyone's thinking is inconsistent to some degree, of course. But how do so many people manage to hold on to a few or several absurd and extreme beliefs while otherwise functioning somewhat reasonably? "We are all experts at applying logic and reason in one area of our lives while judiciously avoiding it in others," warns Dean Buonomano, a UCLA brain researcher.[30] "I know a number of scientists who are unequivocal Darwinists in the lab but full-hearted creationists on Sundays. . . . Because our decisions are the result of a dynamic balance between different stems within the brain— each of which is noisy and subject to different emotional and cognitive biases—we simultaneously inhabit multiple locations along the irrational-rational continuum."

As a member of NASA's Apollo 14 mission in 1971, Edgar Mitchell became the sixth human to walk on the Moon. I met him in 2007 at a space conference in Phoenix, Arizona. We chatted about space exploration past, present, and future. He was sharp and knowledgeable on that topic, as one would expect given his background. However, for all his accomplishments and intelligent demeanor, some might describe Edgar Mitchell as another spoon-bending Uri Geller, only with a better résumé.

In 1973, Mitchell founded the Institute of Noetic Sciences, an organization dedicated to such ventures as researching and educating about "mediumship" (communicating with dead people). Mitchell has personally supported, encouraged, or closely associated himself with belief in UFOs, the Roswell myth, the quantum hologram, telekinesis, hauntings, remote healing, and tarot-card readings—all of which do not stand up to the test of science and logic. In fact, he even claims to have conducted secret ESP experiments with an associate on Earth while he was aboard the command module during his lunar mission.[31]

I am not suggesting that it is always wrong to investigate fringe possibilities. Some wild ideas turn out to be valid. But some of Mitchell's words and works clearly come across as confident affirmations of claims for which no one has ever produced convincing scientific evidence. For example, unless it is being kept secret, neither his nor anyone else's work on the quantum hologram[32] has produced anything of substance. To date there has been no public presentation of evidence or experiments the world's scientists have been able to study, replicate, and confirm.

Fig. 1.1. NASA astronaut Edgar Mitchell has an impressive list of accomplishments to his credit, including walking on the Moon in 1971. He has also been a supporter of belief in UFOs, the Roswell myth, ESP, remote healing, tarot-card readings, and other dubious claims. *Photo courtesy of NASA.*

Why would someone with such a scientific professional background believe these things? Like Tredinnick, Mitchell is highly educated. He earned a bachelor of science degree in industrial management from the Carnegie Institute of Technology, another bachelor of science degree in aeronautical engineering from the US Naval Postgraduate School, and a doctorate of science in aeronautics and astronautics from the Massachusetts Institute of Technology.[33] Those degrees, combined with the challenging and extensive training he underwent to become a naval aviator and Apollo astronaut, make him easily one of the most educated and highly trained people who have ever lived. Nevertheless, he apparently is convinced that many extraordinary and unlikely claims that have not been confirmed by the scientific method are indeed valid anyway.

It's not as if something is wrong with people like Tredinnick and Mitchell. Given the way our brains work, it is likely that they and people like them are normal. It's the skeptics who are behaving weirdly by not passively believing extraordinary popular claims without evidence, such as astrology and ESP. These two men exemplify the inherent gullible nature of humanity. Without the shield of skepticism and critical thinking, any one of us is capable of believing virtually anything—regardless of intelligence and academic credentials.

SO WHAT?

Some people wonder why any of this matters. I hear it all the time. When I give a lecture or take listener calls as a guest on a radio show, inevitably someone asks, "So what? Who's it hurting?" I understand the notion well, as I have considered it myself many times in the past. Who cares if significant numbers of people believe in reincarnation (20 percent), miracles (76 percent), hell (61 percent), and witches with real magical powers (18 percent)?[34] Isn't it their right to be irrational, their personal business if they want to believe in extraordinary things that have not been shown to be real?

There are two key problems with the *who cares?* reaction. First, any skeptic with a conscience has no choice but to care. It would be heartless to sit back and observe a world with so many people stumbling around in the dark and not direct them to the light switch.

Of course people should have the right to believe things within the privacy of their minds. Freedom of thought is a fundamental human right, or should be. But good thinking is a noble goal for all of us, a goal that supports this freedom rather than attacks it. What good is freedom of thought if one doesn't know how to think in the first place? I don't push for others to agree with me on every issue, claim, or belief. I work to inspire and teach others to think for themselves and draw their own conclusions. Second, one ought to care as a matter of self-interest because irrational beliefs impact everyone in one way or another. Irrational beliefs hurt us collectively by acting as a massive drag on society, slowing human progress every moment, everywhere. Indifference about billions of people trusting their health to medical quackery, squandering their money on lies, and looking to superstition for motivation and meaning in their lives equates to not caring about humanity.

It is crucial to recognize that bad thinking in one area of life can bleed over into other areas of life. A cherished bogus belief may be more than a lone mistake. It's often a symptom of a bigger problem. If belief in something as unlikely and unjustified as alien abductions or tarot cards is able to slip by one's cognitive defenses, then many other bad ideas probably will too. This is important because the next false claim to come along may be something more than a charming eccentricity. It might lead to real trouble. For example, there is no good reason to think that sloppy thinkers within democratic societies will suddenly morph into rational beings once they enter a voting booth. Business leaders who don't understand how their own brains work are less likely to realize when biases and fallacies threaten to lead them astray. This is how a CEO ends up making irrational moves like spending millions of dollars on advertising that doesn't work or paying more than $100,000 to have a buffalo head buried on a construction site in order to ward off the threat of mischievous genies—yeah, that happened.[35] The cognitive routines that prod one to accept imagined conspiracies and other fantasies as real don't automatically go away when the stakes rise. We can't trust our propensity to believe the unbelievable to be consistently restrained by imaginary borders that mark off some limited territory reserved for nonsense within our brains. Those same cognitive processes and subconscious biases that allow someone to believe in relatively harmless claims can easily lead that person into irrational decisions about how best to treat a child's serious illness or whether to invest his life savings in a "too-good-to-be-true" business opportunity. Consistency is crucial to good thinking.

TIME WAITS FOR NO BELIEF

Try to imagine the time irrational beliefs have stolen from the one hundred billion or so people who have ever lived. Think of the hours, days, and years squandered on extraordinary claims that turned out not to be true or remain unlikely to be true. This self-inflicted slaughter of time has gone on for millennia and continues right now. It is a massive tragedy that gets no attention. With all the challenges before us, can humankind really afford to burn up so much of this finite resource on dead-end claims and unlikely beliefs?

Let's do some quick math. Imagine a person who believes that psychic readings are valid, that special people really can know a stranger's thoughts, past, and future. If you prefer, substitute a bogus medical treatment or campaigning for some vacuous or dishonest politician.

We can be conservative and say that our hypothetical true believer spends no more than an average of one hour per day on psychic matters. This would include not only time spent participating in psychic readings but also reading books about psychics, watching TV shows about psychics, talking with friends about psychics, visiting psychic websites, thinking about previous psychic readings, and so on. A total of seven hours per week is probably too low for most true believers, but let's go with it. It adds up to 365 hours per year, all devoted to a claim that no one has ever presented any scientifically verifiable evidence for and that good thinking easily tags as a likely scam and/or delusion. What is the time investment over a lifetime? Let's say our believer fell hard for the psychic claim when she was twenty-five and lived to the age of seventy-five. Fifty years of belief activity equates to about 18,250 hours—that's more than 760 full days and nights sacrificed to belief in psychics over her lifetime! That's too much time to sacrifice for a claim that is almost certainly untrue. How might she have used that time more constructively for herself, her family, and the world?

Some might suggest that belief in psychics brought more joy and meaning to her life. But aren't there better ways to find joy and meaning than paying someone to lie to you?

What could you do with an extra 18,000 hours? Spend more time with your children? Pay more attention to your lover or spouse? Have more fun with friends or pets? Earn a college degree? Write a few thousand poems? Maybe you would nap more, cook better meals, try new sports, create more art, or train for an assault on Everest. What summits might billions of people have reached had they been freed up from the time suck of their irrational distractions?

THE PAST WITHIN US

Psychology professor Hank Davis wants everyone to recognize that we are still prehistoric people when it comes to our thought processes and decision making. He writes in his book, *Caveman Logic*:

> In some ways, we are the lucky ones. Our civilization has never been as knowledgeable as it is today. There has never been as much scientific understanding of the world around us as there is right at this moment. We have put many faulty beliefs behind us. Yet, because our minds are no more evolved than they were a thousand years ago, we are just as vulnerable to illusion, comfort, and social pressure as our ancestors were. Faced with incomplete sensory evidence, we are just as likely to come to the wrong conclusion and then fight passionately to maintain those views because of the comfort or stability they provide. We may be less ignorant than our ancestors, but all the hard-won additional information does us no good if we don't let it inform our thinking. . . . The spectacular accomplishments of some of us define the potential of all of us. With training and social support, we can all work around the lure of Caveman Logic.[36]

Naturalist Edward O. Wilson spent most of a long and illustrious career studying ants, but along the way he developed piercingly honest perceptions of our species: "Humanity today is like a waking dreamer, caught between the fantasies of sleep and the chaos of the real world. The mind seeks but cannot find the precise place and hour. We have created a Star Wars civilization, with Stone Age emotions, medieval institutions, and god-like technology.

We thrash about. We are terribly confused by the mere fact of our existence, and a danger to ourselves and to the rest of life."[37]

"What, then, are we to make of ourselves?" asks anthropologist Ian Tattersall.[38]

> Well over three billion years after life established itself on Earth, we, alone among the millions of descendants of our ancient common ancestor, somehow acquired not just a large brain—the Neanderthals had that—but a fully developed mind. This mind is a complex thing, not in the sense that an engineered machine is, with many separate parts working smoothly together in pursuit of a single goal, but in the sense that it is a product of ancient reflexive and emotional components, overlain by a veneer of reason. The human mind is thus not an entirely rational entity, but rather one that is still conditioned by the long evolutionary history of the brain from which it emerges.[39]

If we desire good thinking in the present, we must always attend to the past within us. Our challenge is more than to think now and to look forward. It is also to acknowledge and approach with wisdom those gifts and burdens given us by our brain's evolutionary past.

THE BELIEF BUSINESS

What is the economic impact of bad ideas, lies, and delusions? How big of a dent do they make on an individual, a nation, the world? "Unimaginably colossal" might be a fair guess. The selling of unsubstantiated claims, nonsense, and outright lies is a vast international business—and business is booming. According to one study, Americans alone spend more than $34 billion per year on alternative medicines and therapies, most of which don't work and some of which are dangerous.[40] Some of it probably does some good for some people. But how do we find the good in this massive jungle of alleged treatments that runs the gamut from outright fraud to sincere-but-wrong claims if the scientific process is not the final say?

Many of today's health stores, grocery stores, drug stores, and pet stores offer some of the most absurd products imaginable—and people buy them. It's as if science never happened. For example, did you know that a $15 bottle of the right plant extract will extract loneliness from your life (water violet), end feelings of shame (crab apple), and help your fear of losing control (cherry plum)? At least those are the claims of products for sale in stores and online right now. I even found a pill specifically for "writer's cramp" and another product that promised to treat "despair," despite having no active ingredients listed on the label. Better yet, I spotted the answer to an "unfulfilled life" in the form of a $15 bottle of homeopathic wonder-water. Just a drop or two on your tongue, and fulfillment rushes in, I presume. And did you know that a $30 "energy blueprint" water treatment can stop your dog or cat from bullying other dogs and cats? The product is so potent that pets don't even have to consume it or have it applied to their bodies. Just a few drops somewhere in their living space is enough to compel them to work and play well with others. Those who prefer to keep their quackery close can turn to expensive, copper-infused underwear and socks for treating their aches and pains. No credible evidence or reasons to think any of this stuff helps, of course, but people fall for it nevertheless. In the absence of good thinking, there seems to be no limit to what one person is willing to buy from another.

The sports and fitness industry is also in on this profitable game, selling products so silly that a few seconds of good thinking can deflate with ease. Colorful tape heals injured muscles, even when applied by people with no medical training? A drink with high sugar and sodium content will help athletes win? Bracelets with "holographic technology" improve athletic performance by "reso-nating with your natural energy field"? Power Balance was sucking up mountains of cash from weak skeptics with their bracelets until they were legally forced to state the following: "We admit that there is no credible scientific evidence that supports our claims and therefore we engaged in misleading conduct."[42] But don't weep for Power Balance's demise. Apparently the business is alive and well. Despite the fact that it was widely debunked and the company is no longer able to directly claim that its bracelets actually do anything, they continue to be prominently displayed and sold in sporting-goods department stores and other retail and online shops across the United States. This happens again and again, by the way. A product is exposed for what it is—but people keep lining up to buy it anyway. To give some consumers the benefit of the doubt, one may assume that some of them may not have heard the news about a bogus product. This is why research is an important component of good thinking.

Fig. 1.2. Even pets are not safe from our sloppy thinking. The products shown here don't have anything one could reasonably think of as medicine or even an active ingredient. The *water* in these bottles is supposed to be empowered with the "energy blueprint" of I'm-not-sure-what to stop a cat or a dog from bullying other animals. Another bottle is for pets with self-esteem issues. Here's the manufacturer's explanation of the "Bully Remedy" on its website: "Spirit Essences are not like medicines, drugs, or supplements and need not be swallowed. Once they contact the animal (even slightly via an animal's living area, such as bedding, brush, or spraying inside a kennel or carrier, etc.), they are doing their job. This is because the remedies are energy-based. The ingredients do not contain actual plant, animal, or mineral material, but the energy blueprint of various plants, animals, minerals, objects, or even places."[41] *Photograph by the author.*

It has never been easier to perform a quick check of a product or claim. No matter where you are, thanks to smartphones, enlightenment is often seconds away. The next time you are in a store, about to buy something weird or miraculous, simply do a quick Web search for its name. But don't just do a price comparison or features check, as many do. Search the product name with keywords such as "scam," "skeptic," "controversy," or "fraud." Don't make the mistake of searching for the product name alone because you are likely to get back nothing more than an official website and a string of endorsements if it's popular, as many bogus products are. You want to hear what the doubters are saying. A bad review or condemning skeptical analysis you find on the Internet may not necessarily be accurate, either. But it's important to try to find some of those opposing views because they may get you thinking more clearly about the product that is tempting you.

Regardless of how far removed from the concepts of logic, evidence, and scientific validation some products may be, they will succeed if their pushers manage to push the correct psychological buttons. First of all, packaging matters. If the surface is attractive enough, we may not bother asking ourselves if there is any substance below. Simply being beautiful or interesting, or stimulating a positive emotional response in us, can be enough to make us buy—even if the product is otherwise worthless. I have seen ridiculous products in health-food stores and New Age shops that were so beautifully presented, for a brief moment, I wished I wasn't so sensible so that I could buy them! Go ahead and laugh, but some healing crystals really are beautiful. I once bought a good-luck talisman at a market in Amman, Jordan, not because I thought it had magical properties but because I just liked it—and it was cheap. Something about its design hooked my brain and reeled me in. I currently have two Thai Buddha statues in my home, not because I am a Buddhist, believe in reincarnation, or anything like that. I bought them because they were visually appealing to me. I also bought Buddhist prayer beads while in Nepal. Why? It is natural and common for us to buy things because they make us feel good. Nothing complicated about it, no other reason than we like them. I don't regret those purchases, because I got exactly what I thought I was paying for. But what if I had expected more? What if I thought my good-luck talisman would allow me to win the lottery? Its visual emotional appeal (real) would have conspired with my irrational belief in its magical powers (not real, probably) to seduce me and I would have been vulnerable to severe price gouging. When trying to make reason-based decisions in the marketplace, it helps if we remember how easy the physical beauty of a package, a picture on a label, or a well-crafted phrase can influence us.

HOW TO MAKE HAPPY TALK
WITH IRRATIONAL BELIEVERS

When someone tells me about a ghost in his bedroom last night, the miracle cure that saved her life, or the presence of a god or gods during prayer, I resist reflexively blasting him or her with a lecture on common sensory shenanigans and handing over a list of relevant scientific studies because I know I'm not only talking to one person. I have to reach the conscious person talking to me *and* the subconscious person somewhere within. Both are making the case to me, both are analyzing me, so I need to communicate to both.

The first key is to avoid being perceived as threatening. The shadow brain reacts fast and decisively when a threat is detected. So, in order to avoid having the drawbridge raised and ending all hopes of further meaningful communication, I reassure people early on that I'm not out to demolish their happiness or deny them comfort. I'm nothing more than a humble guy trying to figure out the universe the best I can by questioning everything. I'm there to help, not hurt.

It is important to make a sincere effort to listen. I learn a lot by listening and highly recommend it for everyone. I think it sends a subtle message to the shadow brain scrutinizing me that I'm not an arrogant jerk with an underhanded agenda. This helps keep that drawbridge lowered. I also try to be likable, not in a fake and manipulative way, but in a manner that reveals my true positive intentions. If someone's shadow brain decides that it dislikes you, then it will try to prevent the conscious version of the person you are talking to from liking you. And we find it more difficult to learn from people we dislike. Be nice, be heard.

Over the years I have learned that it can be productive to discuss cases of derailed reasoning that are removed from whatever specific unlikely claim or belief the person has become attached to. For example, if the person is wild for Bigfoot, maybe talk about general problems with belief in the Loch Ness monster. This can keep the conversation from feeling too personal or threatening. Let them figure out the connection on their own.

Finally, to avoid making the encounter unnecessarily confrontational, which is counterproductive in most cases, I try to frame it as "You and Me vs. Delusions/Lies/Errors" rather than "You vs. Me." I work to convey the idea that if we both care about what is real and true, then we are on the same side. The point is not to impress anyone with your superior knowledge or win an argument. The goal is to offer a helping hand to fellow humans and inspire them to embrace good thinking.

In many countries around the world one can watch television programs and infomercials that feature Christian preachers offering tiny vials of "miracle water" or small pieces of what they say is supernaturally infused "prayer cloth." They promise or suggest that possessing these things will heal "every" disease as well as bring great financial rewards to anyone who calls in to request it. Out of curiosity, I've done this, and typically what happens is that you are sent the oil or cloth as promised, but along with it comes a steady stream of letters pleading for donations that continue to come for months and even years in some cases. These preachers apparently figured out that by giving something to people, even worthless junk, they create a sense of obligation in the person, which makes her or him more likely to send money. Judging by the tastes some of these popular religious leaders have for mansions[43] and luxury jets,[44] as well as the high cost of television time, it is reasonable to assume that many people do call and many of them give away their hard-earned money.

I hope no readers will feel it is antagonistic to point out that religion is big business worldwide and that many people are exploited and harmed by it in various ways. This should not be interpreted as condemning all religions and all religious people. Of course many positive things come out of organized religions, and many millions of decent, honest people do wonderful work in the name of their belief. Nonetheless, there certainly is no shortage of things to criticize within religions around the world. And, if one cares about the well-being of religious people, isn't there a moral obligation to speak out? Like anyone else, people who believe in a god or gods need critical thinking and skepticism to protect them from deceits and delusions that can cost them their time, money, and safety.

Good thinking is a powerful defense against many of the problems we see within religions. A little doubt goes a long way toward keeping pious predators away. Without getting into supernatural claims and doctrinal issues, I advise people who give money to churches, mosques, temples, gurus, and the like to make sure they are thinking clearly about where their money is going.

The staggering sums of money pushed around every day in the name of contradictory gods and religious institutions that make conflicting claims can only mean one thing: Huge numbers of people are investing money in claims that are not true. They are wasting their money. If you don't see how, then it's time for a bit of good thinking. One and a half billion Muslims claim there is only one god, and many of them routinely invest some portion of their time and income in this claim. Meanwhile, 1.2 billion Hindus say that there are millions of gods, and many of them routinely invest time and income in this claim. But both of these claims cannot be correct. It's math. There is no middle ground, no compromise here. I have spoken with many Hindus in India and Nepal, and they assured me that their gods are not amorphous entities that could be a multifaceted expression of one god, as some say. No, they are distinct, separate gods, each with a unique personality and biography. I have discussed Islam with Muslims in Egypt, Syria, and Jordan who were exceedingly clear about their god being "one" and definitely not some combined entity drawn from multiple gods. So there seems to be no logical reconciliation possible between these particular beliefs in one god and many gods. Both claims cannot be true. Maybe one is true and one is false, or maybe both are wrong. If we can agree on this, then there is no way around the obvious: many hundreds of millions of people are wasting a lot of money by funding and supporting false claims. In addition to the Muslims and Hindus, we have more than two billion Christians, most of whom are absolutely certain that Muslims and Hindus have it all wrong. Many Christians, of course, routinely invest portions of their income into their religion.

In the United States religious organizations, most of them Christian, are granted tax-exempt status, which amounts to approximately $71 billion in lost revenue for the government each year, according to one analysis.[45] This is a huge sum of money—more than a trillion dollars every fifteen years. Although there is much doctrinal overlap, there are a significant number of Christian organizations in disagreement with one another over irreconcilable claims. Logically this means at least some of these organizations are avoiding taxation via false claims. I'm not challenging anyone's sincerity, but, again, everyone can't be right. Hopefully even the most religious reader can recognize that caution and a rethink are warranted at this point. No matter what some religious leaders may say, good thinking should never be muted or abandoned at a religion's doorstep. Besides, why would any gods worth our allegiance be opposed to a bit of constructive skepticism and the vigorous application of a human brain?

THE IRRATIONAL READER

It is too easy these days to point to the mind-numbing distractions and social rot brought on by the worst of television and the World Wide Web. But what about books and reading? What about nonfiction books in particular? Pulitzer Prize–winning journalist Chris Hedges cites several disturbing statistics in his book, *Empire of Illusion: The End of Literacy and the Triumph of the Spectacle*. Here are two that stopped me cold: After leaving school, one-third of US high school graduates never read another book for the rest of their lives, and 42 percent of college graduates never do again.[46] Here is how Hedges bemoans the state of American literacy:

> We are a culture that has been denied, or has passively given up, the linguistic and intellectual tools to cope with complexity, to separate illusion from reality. We have traded the printed word for the gleaming image. Public rhetoric has been designed to be comprehensible to a ten-year-old child or an adult with a sixth-grade reading level. Most of us speak at this level, are entertained and think at this level. We have transformed our culture into a vast replica of Pinocchio's Pleasure Island, where boys were lured with the promise of no school and endless fun. They were all, however, turned into donkeys—the symbol, in Italian culture, of ignorance and stupidity.[47]

Nevertheless, books continue to matter to some degree, and those who do make the effort to read nonfiction books can be viewed as at least making an effort to better their minds, certainly more so than those who don't bother. But can we assume this? Perhaps not in light of what passes for nonfiction these days. I am a lifelong lover of books. I think there is something very special about "the book." It is an extraordinary vehicle for ideas, questions, and dreams that can outlive its author. Books are my kind of magic. But not every book inspires me with such positive feelings. Wander through any typical bookstore today, and you are likely to find yourself on a disorienting tour of lies and lunacy. Browse through the offerings of an online bookseller, and you are sure to be alarmed by the amount of barely camouflaged refuse available. If there is a limit to it, I have yet to find it. Want to be a time traveler? If so, there is a "nonfiction" book waiting to show you how. Fancy becoming rich and famous this year? I found a book that will get you there. All you have to do is believe.

Books classified as nonfiction promise to reveal to readers ways they can connect with the "other side" and chat with dead people. Other books describe "cryptids," mythical creatures of all shapes and sizes but with one thing in common—they probably don't exist. The authors of these books, however, seem to know a lot about them, in great detail. Want to lose fat and get fit? Plenty of books have been written with you in mind. Only be careful which one you choose, because many authors care little if at all about science and reality when it comes to diet and exercise. If it's meaning and purpose you seek, many books await you. They feature New Age gurus pushing old-age claims that are not supported by good evidence. Some of these authors hijack the concepts and vocabulary of science, especially physics these days, in an attempt to seem sophisticated and take on the scent of real, cutting-edge science. Sadly, it works on millions of people. They read a few sentences laced with the words "quantum" and "cosmic" and are sold on whatever unsubstantiated claims the author makes.

Social-science sections are intellectual minefields for the weak skeptic. Venture in without your good thinking up and running, and you will be wounded. Attractive and authoritative "nonfiction" books tout conspiracy theories and interpretations of historical events that travel well beyond credible evidence and facts. "We in the West are currently going through a period of fashionable conspiracism," writes David Aaronovitch in his book, *Voodoo Histories*.[48] "Books alleging secret plots appear on the current affairs and history shelves as though they were scholarly or reliable as works by major historians or noted academics. Little distinction is made between a painstakingly constructed biography of John F. Kennedy and an expensive new tome arguing—forty-three years after the event—that the president was killed by the mafia."

Politics is a fertile field for lies and nonsense. No surprise there, but something about the implied credibility of a published book can make the worst political monsters and jesters seem almost human. Hardback hatchet jobs on political figures or campaign fluff spread across three hundred pages are case studies in dishonesty and confirmation bias, yet people snatch them up by the millions. Among the self-help books, one finds an abundance of medical quackery catering to literate sloppy thinkers. No matter your ailment or your fears, there is a book promising to cure you. So who buys, reads, and believes all this stuff? Those who fail to practice good thinking, that's who.

In recent years, several books written by people who claim to have taken a roundtrip visit to heaven have earned millions of dollars. These books are marketed as accurate accounts of real experiences, shelved and sold as nonfiction works. *Heaven Is for Real* sold more than ten million copies and inspired a film that grossed more than $100 million.[49] *The Boy Who Came Back from Heaven* was a *New York Times* bestseller that sold over a million copies.[50] *90 Minutes in Heaven* (not to be confused with the book *23 Minutes in Hell*) sold more than six million copies and was on the *New York Times* bestseller list for more than five years.[51] I read *90 Minutes in Heaven* and have skimmed through several others in this unique genre. What I discovered is that—surprise—these are stories and nothing more. Good thinkers know that extraordinary tales without verifiable evidence are just not good enough when the claims they make are unusual or important. An important story shouldn't be automatically ignored; it might be true, but it's at best a starting point. Good thinking demands more. Any thoughtful skeptic would recognize immediately that all of these heaven tourism accounts are not to be trusted without evidence. To be clear, I don't pretend to know that these authors did not visit heaven, and I see no need to question their honesty or sanity. It is enough to know that such personally profound events are common to the human experience and can feel completely convincing to the person who goes through them—even though it all takes place within the confines of his brain. No doubt there may have been some kind of an emotional and overwhelming experience for some of these authors. But for the rest of us, such stories are nothing to get excited about and certainly are not worth believing in without evidence. The reason so many people eagerly accept these stories as true is because they want to believe them. For many, the claims match perfectly with their preexisting beliefs, so they fail to doubt, to ask questions, or to request evidence. Such gullibility is the antithesis of good thinking. It is worth mentioning that the credibility of one of these books, *The Boy Who Came Back from Heaven*, crash-landed for all to see in 2015 when the boy, Alex Malarkey, admitted that he made up his story to get attention and never went to heaven.[52]

One of the all-time great efforts in credulity harvesting via the printed word is *Chariots of the Gods?* This classic work of pseudoscience written by former hotel manager Erik von Däniken became an international bestseller and still inspires people to believe in ancient alien astronauts more than forty-five years after it was first published in 1968. It doesn't seem to matter that the world's professional archaeologists have loudly and repeatedly condemned the book and provided thorough explanations for why its claims are hollow. Yes, ancient people were smart enough and capable enough to build pyramids and other large structures on their own with no assistance from aliens, as archaeologists have shown. No, tribal art does not prove alien visitations no matter how Von Däniken interprets it. For a skeptical analysis of the *Chariots of the Gods?* claim, see my book *50 Popular Beliefs That People Think Are True*, which includes a chapter about it.[53]

Chariots of the Gods? and other books like it provide us with a glaring example of why learning *how* to think is so important. We can't expect to hear a specific warning for every bogus book or product for sale, because there are too many and crooks and crackpots producing new ones every day. Most people are going to go through their lives without ever hearing or reading anything from the archaeology community about Von Däniken's work. But that won't matter for those who are good thinkers and are able to adapt to whatever intellectual challenge they happen to find in front of them. All on their own, people can work through whatever extraordinary, weird, dangerous, or important claim comes along. Good thinkers will know to give it a pass when the evidence doesn't match the sales pitch.

As a side note, it is interesting that the title of Von Däniken's book included a question mark when the book was originally published in 1968. But *Chariots of the Gods?* became *Chariots of the Gods* somewhere along the way. Why? Could it be that the publisher or author assumed that people were more sensible in the 1960s and 1970s and therefore felt the need to present the title in the form of a question? Today, however, perhaps so much of the reading public is gullible enough that a declaration for the title is considered safe. Perhaps this makes sense. As I write this, *Chariots of the Gods* is ranked sixteenth (based on sales) in Amazon's "astronomy and space science" category, perched ahead of nonfiction books by many prominent scientists, including Stephen Hawking, Carl Sagan, Michio Kaku, Brian Greene, and Neil deGrasse Tyson.

Julien Musolino, an associate professor of cognitive psychology and the director of the Psycholinguistics Laboratory at Rutgers University, recognizes the need for more good thinking in society:

> As someone who has been teaching at the university level for close to two decades, I can tell you that poor critical thinking is endemic. Last semester I had an undergraduate student approach me and tell me, after he explained to me that he was a science major, that he did not believe in human evolution. Much more revealing perhaps, were the arguments he invoked, chief among them the claim that "nobody has ever seen human evolution happen" and that science is all about "things that can be directly observed in reproducible experiments." I kindly told him that it was a safe bet that nobody saw his parents have sex, but that we could rather reliably infer that it did happen by his mere presence in the room. That actually got him to think.[54]

Musolino recalls a time he was pulled into a debate over the age of the Earth with about a third of the students in an undergraduate seminar he taught. He says he tried his best to convince them that its age is not a matter of opinion but a scientific conclusion based on many lines of evidence.

"These are of course anecdotes," said Musolino, "but research on scientific literacy in the United States paints very much the same picture, as do national polls designed to gage the general public's level of rationality. The fascinating research by Frank Keil on 'the illusion of explanatory depth' is also relevant here. Extreme levels of religiosity, denial of evolution and climate change, or the effectiveness of vaccines, to mention a few of the most infamous cases—there are many others—are also very revealing in this regard. The reason we should worry about this is that the US is rich and powerful, and thus extremely influential around the world.

Moreover, those of us who live in the US can try to do something to combat this problem from within."[55]

It should be easy to recognize the potential problems Musolino warns of when large numbers of educated people can't or won't accept even the most basic of scientific facts. Pleas to "save the Earth" and calls to end war and poverty, for example, seem unlikely or far-fetched given that many of us do not even see the world as it is now because of bad thinking. How much more difficult must it be to solve a big problem when it is approached from a position of self-inflicted confusion and delusion?

DIVIDED THINKING

Based on current evidence, people of one kind or another have been around for at least two million years, and anatomically modern humans like us came on the scene only about 200,000 years ago. Yet in this relatively brief span of time, we have achieved great and improbable things. Our ancestors tamed fire in the wild and then learned how to create it. We used those precious and powerful early flames to light up the night, cook food, and turn back the predators who had long hunted us. With the creative power of our brains, we shaped common sticks and stones into uncommon tools. We expanded throughout Africa and beyond to colonize almost every environment on Earth. We were the ones who delivered art, music, language, mathematics, science, and civilization to the planet.

This story of unlikely survival and intellectual/technological advancement among so many other life-forms makes a convincing case that—at least in this small corner of the universe—we are extraordinarily special creatures. Even with our many flaws and limitations, we stand out in positive ways. And despite the long shadow of so much self-hate and self-destruction, there is no denying the endless streams of love and cooperation that also define our species. Optimism is not irrational. I have interviewed survivors of the Nazi Holocaust, the atomic bombing of Hiroshima, and Rwandan Genocide who still had love in their eyes and hope in their words. Shortsighted impulsive destroyers of life we may be, but we also appreciate and protect life to a degree far beyond any other known species. We may have imagined and built rockets to murder our neighbors, but we also send them to other worlds to uplift ourselves through the discovery of new knowledge. Clearly there is something in us worth building on. Our potential for good is obvious. What if there were a way to profoundly improve ourselves, to be less prone to fear, hate, and violence? There is.

Near-universal good thinking may seem like a far-fetched dream from the perspective of those of us marooned here in the early twenty-first century. But isn't it a worthy goal, one worth working for? Consider the root causes of so much hate and violence today. *My god is better than your god. My people are better than your people. My birthplace is better than your birthplace.* So much of the unnecessary harm we conjure up on a daily basis is the direct result of irrational decisions made by people in positions of influence and by the rest of us in moments of influence. Avalanches of avoidable problems will never cease tumbling down on us so long as billions give uncritical allegiance to baseless causes, ideas, traditions, and claims. We will forever struggle with mass hate and murder if billions continue to follow incompetent, dishonest, or crazed leaders. Too much of the world is a dark, bleeding mess. Now is the time for good thinking so that our future might be better.

Chapter 2

WHERE DO BRAINS COME FROM?

"You can't see the universe clearly until you know who you are."
　　　　　　　—Alexander Jablokov, *The Place of No Shadows*

Y ou and I know that our thoughts, ideas, emotions, dreams, and memories originate within the brain. But we ought to be humble about this awareness and thank science because for most of human existence, it was not at all obvious. Only the painstaking process of observations and experimentation, carried out by smart and curious people, revealed it to us. The most extraordinary organ of all is not so extraordinary in appearance. It does not glow with obvious importance when compared with kidneys, livers, and lungs. It does not radiate its central role to human life in a manner that is plain to see. The basic knowledge that the brain is where thinking takes place eluded us for thousands of years.

The first known attempts to locate thoughts, emotions, and memories in the human body missed the mark, to say the least. The brain kept its secrets for a long time. Many smart people, for example, were confident that the heart was a far more likely candidate to be the supreme organ of intellect and emotion. This is why, even though we know better today, the heart remains the symbolic location of love and willpower. For all their concern about preserving bodies and placing internal organs in canopic jars for availability in the afterlife, ancient Egyptians seemed to care little or nothing about the brain. During the mummification process, they routinely yanked it out, bit by bit, via the nasal cavities with a hook and then presumably tossed it out as garbage. There are no known records of human brains being preserved by the ancient Egyptians, and none have ever been found in their tombs.[1]

Intellectual giant Aristotle concluded that the brain was some kind of an organic radiator for cooling the blood as it circulated.[2] Some of the brightest minds of the past thought that the brain was essentially made of sperm.[3] Included was no less a thinker than Leonardo da Vinci, who sketched a dedicated tube in the spinal cord for the transport of semen from the brain to the penis.[4] This sperm-brain belief was popular in East Asia as well. Taoists of the past, for example, claimed it was wise for a man to conserve his sperm so that it might make its way back up to the brain.[5] How sad for all those people of past centuries who lived out their lives never having the opportunity to learn much, if any, true information about the brain's origin, structure, and function. How wonderful it is for us to be alive in a time when so much is known and available for us to learn.

An adult female brain typically weighs about two and a half pounds, male brains three pounds or so. That equates to just 2 percent of bodyweight. An average brain measures only six or seven inches from bow to stern and has a volume of around 1100 to 1200 cubic centimeters. Think of a small cantaloupe to help visualize its size. Despite its workload, the brain runs on a mere twelve watts of electrical power—much less than it takes to illuminate a single 60-watt light bulb. But such stats do not come close to describing the importance and abilities of this organ, of course. As mentioned in the previous chapter, the human brain can far exceed its physical limits by the thoughts it produces. It can conjure up ideas and images of our star system, our galaxy, all galaxies, and even multiple universes laid out, end to end. The brain is the finite and temporary machine capable of glimpsing forever. If you were given an option of somehow surviving with just one part of you left intact, the choice would be clear. The brain is you, the rest a supporting cast, mere plumbing. Nothing else comes close to its value. Nothing else makes us human in a meaningful way. With the brain, we can contemplate tomorrow, next year, and even the time beyond death and extinction. The human brain cures diseases and turns letters into poetry. It kills. It saves. It hates. It loves. The brain is everything.

While writing words for this page, my brain imposed an unscheduled break on me. It transported me to a clump of bushes outside my window. Suddenly I was inside a strange, uncluttered place—the brain of a small lizard. I felt only fear and hunger in there. Nothing else existed. Nothing else mattered. The universe had become the branch I gripped. My reptilian eyes darted around in a high-speed effort to spot predator or prey. I knew that my eyes had to see them before they saw me. An ant wandered by and, without a thought, I pounced and snatched it into my mouth. For the briefest of moments, I might have felt a slight tingle of something like satisfaction or pleasure, but I'm not sure. I resumed my frantic scanning for another ant, but then a feeling of electrified anxiety surged through me. Only a moment later did I realize that the fast-growing shadow around me was that of a bird rapidly descending. I ran as fast as my four legs would take me, but the shadow grew larger still.

Fortunately, my brain spared me the experience of being eaten alive. It took me back through the window and didn't stop until I was safely inside my own head again. On the way back to my brain, I slowed just enough while shrinking to see bone, membranes, fluids, tissue, cells, molecules go by. When I finally stopped, I found myself staring across a vast, dark expanse. There was no end to it, at least none I could see. I flew—or swam, not sure—until I came upon several huge, glowing spherical clouds. I think they might have been electron shells. I stopped all movement and drifted in the silence for a moment so that I could try to take it in. Then I found myself writing again. All that, and I never even left my desk. Don't make the mistake of forgetting or never caring about your brain's ability to dream without limits and take you anywhere. It's the best part of being human.

Given what the brain can do, what it means to our personal existence, humankind's long-term survival, as well as every achievement to date and to come, one would think that everyone on Earth would value it and want to learn as much as possible about what it looks like, how it is put together, and how it works. Yet most people know little, if anything, about the brain's origin, anatomy, function, and needs. A few centuries ago, such ignorance was understandable because nobody on the entire planet knew much about those things. In the twenty-first century, however—excluding the many millions of us currently enduring extreme poverty—what excuse is there for anyone not to learn about the brain? Why don't more people care enough to discover and understand the one thing that makes them a person and something more interesting than, say, a rock? Probable reasons for the lack of interest stem from the hidden and complex nature of the brain. We can't easily see it like we can our skin, eyes, teeth, fingers, and so on. In this case, out of sight means literally out of mind, I suppose.

Because the brain is more complex than *anything else in the known universe*, it is perhaps understandable that many people may find it intimidating or at least less inviting to research and contemplate than, say, fantasy football. The truth is, however, that the brain is not beyond the reach of anyone with a desire to learn about it. This is important knowledge because good thinking requires more than skepticism and critical-thinking skills. We also have to know something about how our brains are put together and how they work. Imagine how limited a Formula One race car driver's ability would be without at least a fundamental understanding of the car's engine. In the same way, our ability to reason and make sensible decisions is limited if we do not possess fundamental knowledge of the brain. Don't fail to learn about the one thing that makes all learning possible.

WHERE DO BRAINS COME FROM?

For all the praise I may heap upon the human brain throughout this book, it's important to make clear that it should not be thought of in total isolation from other life. Yes, our brains are the best we've seen at thinking and creating, but do not imagine that they are fundamentally different from chimp brains, dolphin brains, rat brains, and so on. We have a size advantage over some species but not all. We share brain parts and brain systems with many other animals. Our brains are so different in intellectual power because somewhere along the way, evolution selected for the genetic traits that produce it. Although a giraffe's neck may seem a little bizarre to us, we recognize that it's still just a very long mammal neck. Many people make the mistake of thinking there is something profoundly different about our physical brains in comparison with the rest of the animal kingdom. Our brains are the best thinking machines nature has produced on this planet, no doubt, but they are still mammal brains, primate brains to be specific. It is important that we recognize this in order to better understand the prehistoric origins and current limitations of our brains. The truth is, for all their power and glory, human brains are a mess, as any brain researcher will tell you. They have come to us today with a long list of quirks, frailties, inefficiencies, limitations, and troublesome systems that were set in place by the pressures of the long and twisting path of a unique evolutionary history.

To where does the best current evidence point as the origin of the human brain? This is either a very good question or a very bad question. Maybe it is both. We should wonder about our brain's origin and try to know it, of course. But maybe this is a question that cannot be answered in a way most would like it to be because the "origin" we seek is spread across too much time. There is nothing definitive we can identify and say, "there it is." For example, a good starting point might be *Fuxianhuia protensa*, an arthropod in a class of life called Megacheira. Thanks to a *Fuxianhuia protensa* fossil discovered in China, we know that these clawed arthropods lived 520 million years ago and had a tiny but distinct brain.[6] It is tempting to say they possessed the first brain. But we can't, of course. Megacheirans had ancestors, too, and they likely had a brain or something brainlike at least. How does one pick a single point in a seamless march of life and call it the beginning, anyway? How far back do we go? Perhaps we should date the human brain all the way back to the creation of elements within stars more than thirteen billion years ago. After all, there could never have been a brain without those atoms. But that's too far; it feels like cheating. What about the first appearance of organic molecules? Or the arrival of the first entity most people would agree to call "life"? It or they may not have had brains, but I suppose one could argue they represent the most important step toward the brain. Archaea? Bacteria? The earliest multicellular creatures? Maybe the first nerve cell was the first brain. Or the first two nerve cells that established a connection. Was that a brain? What about the first neural net utilized by something like a sea jelly. That's certainly *brainlike*. Or do we have to wait for the first vertebrate with a cluster of nerve cells in its head? The reason it is difficult to point to a sensible place and time for the start of anything is because—apart from extinction—life doesn't provide us with enough convenient punctuation points. It's hard enough just to find and identify fossils that date back millions and billions of years ago. Determining whether any one of them is a specific starting point for any particular trait or feature is nearly impossible.

Evolution is the long blur, a constant, living flow of branching relationships. One thing is always connected to another and another. Having no regard for our love of labels and tidy organization, life rolls on as a continual stream of organic matter. The important thing about the origin of the human brain is not pinpointing some specific time, event, or fossil to declare a beginning in order to satisfy our desire for order. What matters is that we understand the process from which it emerged and how deeply rooted the modern human brain is to its past.

BRAINS FROM BEYOND

Seth Shostak, senior astronomer for SETI (Search for Extraterrestrial Intelligence), headquartered in Mountain View, California, has devoted decades to thinking about the possibility of alien civilizations. If they exist, where are they? What do they look like? How is their culture structured? I asked Shostak to speculate about alien brains. What might they be like?

We think we're the bee's knees, brainwise—the crown of intellectual achievement. But let's face it: our brains are the product of a few billion years of bottom-up evolution, and they barely work. They're not bad in environments, such as the African savannahs, where they spent the longest time being winnowed into their current arrangements. But it's hubris of the first order to believe that our crania are especially good at cogitation. This fact will likely become obvious to us midcentury, when we develop strong artificial intelligence. Indeed, the path from microbe to mankind to machine seems so probable, I feel it's safe to claim that the majority of the intelligence in the cosmos is of the artificial variety.

In other words, if we pick up a signal from E.T., I won't expect that what's behind the microphone on the other end is a biological being. It will be a machine with a better brain than our own—one that's top-down engineered, and one that can greatly improve its successors.[7]

It is necessary to grasp the humble evolutionary origins and resultant limitations of the human brain if we are to have any chance of making the best use of it. Even as one admires it, or stands in awe of it, it is important to recognize and admit that the brain falls well short of perfection in its design and function. A realistic perception of ourselves is crucial. Excessive confidence or arrogance is death to good thinking. In his book *Kluge: The Haphazard Construction of the Human Mind*, New York University psychology professor Gary Marcus calls the human brain a "kluge." This is an engineering term that means a "clumsy or inelegant—yet surprisingly effective—solution to a problem."[8] The name fits our brain well.

"Nature is prone to making kluges," writes Marcus, "because it doesn't 'care' whether its products are perfect or elegant. If something works, it spreads. If it doesn't work, it dies out. Genes that lead to successful outcomes tend to propagate; genes that produce creatures that can't cut it tend to fade away; all else is metaphor. Adequacy, not beauty, is the name of the game."[9]

Understanding Marcus's basic point is the beginning. From there we can move forward. We can praise the brain for what it can do while always acknowledging what it can't do because of its rough-and-tumble evolutionary history.

During a memorable week spent exploring the Galápagos Islands, I thought a lot about evolution, as one naturally does in the place Charles Darwin famously visited in 1835. One morning I sat on a rocky beach as far apart from my group as the guide would allow and imagined Darwin's eyes widening with excitement at the sight of marine iguanas like the fifty or so around me. They sunned on rocks and plunged into the surf to feed on algae and seaweed. Once back on the rocks, they snorted saltwater out of their nostrils while paying no attention to me. Throughout my week on those islands, every sleeping sea lion in the sand, every plant, and every bird above reminded me how fortunate I am to have been born into an age when so much is known about life. Scientific knowledge allows me to appreciate it so much more. Compared to what remains to be learned, I know little, of course, but I try. I am closer to the life that exists today and the life that has come and gone, as a result of awareness, curiosity, and a desire to know more. Anyone can have this. Everyone should have this. Given the brains we all have, it is only human to wonder and learn.

It saddens me that many people around the world reject Darwin's work without ever really understanding it. Some even go so far to think of him as having been an evil liar. They charge him with attacking religion and degrading humanity. The truth is, Charles Darwin did not create or invent evolution. He merely opened his eyes and his mind to discover what was always there.

It may make the complex topic of evolution more digestible for some to point out that it can be reduced to a simple description: *Life changes over time*. That's the heart of what one needs to accept in order to grasp what is the foundation of modern biological science. *Change over time*. Heritable traits, or gene-based characteristics, that are advantageous because they work well in a current environment tend to increase in frequency and reshape a species over time. Sometimes the change is insignificant and sometimes it is profound. Given millions of species and billions of years, it is not so difficult to imagine how our planet's stunning biodiversity came to be. Natural selection and genetic mutation are the dominant drivers of change that can lead to startlingly different outcomes. Change may result in extinction, or in the human brain in an ape that had descended from the branches and stood up. Anyone who can accept the obvious—that life changes over time—should have no problem with the fact of evolution. It's also useful to keep in mind that modern biology is not something to "believe in" or align with like a religion or political party. It is the sum of many discoveries and reasonable conclusions based on the best evidence currently available. Some people say they can't imagine how evolution could work. I can't imagine how it could not work. How would it be possible for all life to stay the same, given so many interactions and environmental changes across such vast stretches of time?

I have discussed human evolution with many people with diverse perspectives and found that a common point of contention is the nature of the brain. Because the human brain is so good at thinking and creating, many people say they can't imagine that it could have just randomly happened. But it didn't "just happen." The human brain exists today only after a long and incredibly complex interaction between many different environments and countless millions of species. I would not describe anything coming out of a multibillion-year process involving literally hundreds of trillions of individual life-forms to have "just happened." Moreover, the human brain didn't have to emerge, and almost didn't. There is no reason to conclude with certainty that the modern human brain was inevitable. One can easily imagine a world and a universe without us.

There were moments in our past when we numbered fewer than a few thousand. Quick extinction would have seemed likely, if not inevitable, to any impartial observer passing by. In fact, there was a world without humans for all but a slim fraction of Earth's history. And the universe managed without us for more than 13.8 billion years. The late paleontologist Stephen Jay Gould described our fragile origin as a "tiny twig on an improbable branch of a contingent limb on a fortunate tree."[10] Nevertheless, here we are. And now things are really beginning to get interesting.

The previous century saw many key discoveries about our brain, but this new century is even more promising. As usual, more knowledge leads to more questions. I encourage readers to keep a close eye on news coming out of brain research and human evolution in the coming years because it's a safe bet that profound things will be learned soon. In the meantime, however, people should not let unanswered questions lead them to reject *answered* questions. The fact that we don't know everything about human evolution is motivation to keep working, not cause to dismiss what we do know about it. Our remaining ignorance about the brain's origins, complexity, and abilities certainly does not justify throwing up our hands in intellectual surrender so that we can pretend to know that our unique intelligence is owed to magic, gods, or ancient alien astronauts.

It will be brains just like yours that will eventually explain away the mysteries that currently frustrate us. Be patient. Those who point to temporary ignorance and claim permanent victory have given up. Don't join that parade to nowhere. Have more respect for the unlikely survivor in your head. "It has often and confidently been asserted, that man's origin can never be known," Darwin wrote in *The Descent of Man*, "but ignorance more frequently begets confidence than does knowledge: it is those who know little, and not those who know much, who so positively assert that this or that problem will never be solved by science."[11]

A BIG QUESTION

Why us? How did we end up being so intelligent? Why are we planning sustainable cities and Mars missions when the rest of the animal kingdom seems content to just make it through one more day? Out of

all the life-forms on Earth, why us? Of the mammals, the other primates, and the extinct human species, why *Homo sapiens sapiens*?
Our brains have tripled in size over the last seven million years, and
most of that expansion occurred in just the last two million years.[12]
Two million years may seem like a long time to you and me, but in
the big picture that's fast. Based on fossil evidence, our ancestors
experienced a huge leap in brain size somewhere between 2 and 1.5
million years ago. Then another jump occurred between 500,000 and
200,000 years ago.[13] In case you are wondering how we know this
when the soft tissue of brains doesn't fossilize, paleoanthropologists
can extrapolate brain size and sometimes even brain features by analyzing the interior surface of fossilized craniums.

This change was not only about bigger brains, however. The
hominin brain was becoming more complex, too. Evolution was
selecting for higher thought, more abstract thinking and, of course,
language. We had transitioned from almost purely instinctual creatures to reasoning, planning people who increased their chances
of survival and reproduction by *thinking* their way through challenges and then sharing solutions and discoveries with one another.
Once we had evolved big brains, something else unusual happened.
Around 50,000 years ago, there was an explosion of cultural behaviors that we recognize today as purely human. The coming years
will be exciting as archaeologists, paleoanthropologists, and brain
scientists work to fill in many of the gaps in our understanding of
the human story.

"Humanity has survived by mind over matter; we have flourished
despite our anatomy rather than due to it," said Portland State University anthropologist Cameron M. Smith.[14] "Early in the evolution
of the genus *Homo*, nearly two million years ago, the mind underwent several evolutionary transitions as we went from using the
body to using the mind as our means of adaptation. These transitions
allowed the mind to expand its 'bubble' of space and time awareness,
among other things. More transitions came later, resulting in modern
mental complexity, significantly involved with the origins of language
and symbolism, not too long after 100,000 years ago."

It helped that as primates we were in the right biological club.
Accounting for body-size ratios, mammal brains are the largest of
all animals', and primate brains are generally twice the size of other
mammals'.[15] But again, don't be misled into thinking size was the

only factor. It was our brain's *reorganization* and *changes in proportional sizes* within that seemed to be most important to our eventual spectacular intelligence.[16] Anthropologist Ian Tattersall, curator emeritus of the American Museum of Natural History in New York City, has a perspective that is typical of many who research the evolutionary path of the human brain. It's not all that special, they say, yet somehow it ended up being profoundly special. He writes in his book, *Becoming Human*:

> The human brain, whatever its marvels, probably does not contain any completely new structures—any structures, indeed, that are not shared with all of our primate—or even mammal—relatives, however humble. Thus we cannot look merely to entirely novel brain components to explain our cognitive powers, however elegant an explanation that would be. What has happened over our evolutionary history, however, is that certain parts of the human brain have become enlarged or reduced relative to others and the connections between them modified or enhanced. Even this is not unique to us, though: for while we undeniably have the largest primate cerebral cortexes (about 76 percent of our large brain's total weight), there has been a dramatic increase in the percentage of the brain occupied by the cerebral cortex and supporting structures among higher primates in general.[17]

Okay, so what we have are essentially rat brains that evolved to work really, really well. But why? How? What was the specific catalyst for the brain's relatively rapid rise to intellectual superiority, at least on this planet? Something we did, or something that happened to us, resulted in this amazing brain of ours, the one that surpassed them all.

The Standing Brain

One of the critical changes linked to our remarkable brains is bipedalism. Standing up did wonders for us some four million years ago in Africa. It freed the hands, for example, allowing our ancestors to carry things and more easily make and use tools. Being taller allows one to see farther as well, which would be an important survival adaptation given the constant danger from predators in Africa at the time. Seeing farther could have led to more roaming and exploring, too. In short, it could have meant significantly more mental stimulation. Standing upright also reduced the surface area of a body exposed to direct sun, making it easier to cool the brain. Bipedalism enabled us to become exceptionally good walkers and runners. We may not be as fast as cheetahs and gazelles, but we do have the remarkable ability to walk/run literally all day long. Modern humans and our human ancestors are/were among the best endurance athletes in the animal kingdom. In fact, a common hunting strategy used by modern-day hunter-gathering societies, and most likely prehistoric people, is to chase and stay within sight of an antelope or some other animal for many hours until it collapses from exhaustion.

For all its benefits, however, bipedalism doesn't seem to be the singular key that unlocked the explosive potential of the primate brain, because nothing much happened to our ancestors' brains for a couple of million years after they stood up.

The Fireside Brain

Was it a hot flame that sparked our brain's rapid rise? Perhaps the sequential accomplishments of observing and studying wildfire, exploiting it, capturing it, taming it, and finally creating it were the keys to the radical expansion and reorganization of the human brain. Fire was a big deal, to say the least, and it is important to appreciate its possible central role in our evolution.

Perhaps two million years ago or even earlier, depending on which anthropologist you ask, our ancestors began the practice of capturing wildfire and using it to increase the odds of survival and improve their lives. Fire enabled them to cook food they had trapped, hunted, and scavenged. This expanded the hominid menu and made meals more nutritionally potent. "Cooking is what makes the human diet 'human,' and the most logical explanation for the advances in brain and body size over our ape ancestors," concludes Richard Wrangham, as quoted in an article in the *Harvard Gazette*. Wrangham, a Harvard University primatologist and author of *Catching Fire: How Cooking Made Us Human*, continues, "It's hard to imagine the leap to *Homo erectus* without cooking's nutritional benefits. To this day, cooking continues in every known human society. We are biologically adapted to cook food. It's part of who we are and affects us in every way you can imagine: biologically, anatomically, socially."[18]

It's probably no coincidence that *Homo erectus* was making good use of fire and also happened to be the first people to spread across much of the planet into many diverse and challenging environments. Fire gave them an edge over predators and allowed them to be more intellectually productive at night. Sunset no longer meant only darkness and danger. Now people could gather together around a fire at night, illuminated and relatively safe, to look into one another's eyes and interact. There was more time, more opportunity to tinker with tools, imagine, and make plans. And we can't underestimate the impact of language on the brain. I don't know about you, but if I'm sitting around a fire, night after night, with my *Homo erectus* buddies, I want some conversation. Humankind's first awkward silence may have occurred at the first campfire gathering. The impulse to communicate more and in greater detail must have been significant and likely led to the development of language, which would have been a major advantage for any group as well as pressure for the brain to change.

The Seafood Hypothesis

Maybe it was dining by the sea that did it. An interesting idea put forward in recent years points to the consumption of mollusks, crustaceans, fish, and other coastal food, as well as aquatic food taken from freshwater lakes and rivers. Waterfront property may have been more valuable in prehistoric times than it is now, given the abundance of seafood that was available to people. This food source could have provided some of our ancestors with the necessary high-quality calories to jumpstart and sustain growing brains. Seafood also happens to be high in the potent nutrient DHA (docosahexaenoic acid/omega 3 fatty acid) which is essential to brain structure and health.[19]

Did Brains Make Tools or Did Tools Make Brains?

I'm not very good at it, but I sometimes attempt to make stone tools. I'm not entirely sure why. Mostly I have learned that it's not as easy as it looks, and finger injuries are not uncommon. I thought I had been originally attracted to the idea of tool making because I imagined having a few faux-prehistoric hand axes and choppers lying around would fit well with the decor of my "man cave." But, as I often tell my wife and kids, I also do it because it puts me in touch with my inner *Homo habilis*. I have jokingly described to my family that it feels good to activate dormant genetic memories of my distant ancestors by doing what they did. Stone tools made us, I declare, so I make stone tools. But maybe it's no joke, after all. Maybe I'm onto something.

Some scientists today think early tool making drove those bursts of development in our brain's evolution. They think that inventing and making tools put us on the cerebral fast track more than two million years ago. Crafting stone tools, particularly using one to make another, requires imagination, foresight, and hierarchical thinking. It's not easy to make good stone tools—I can vouch for that. If being able to make them was an advantage in prehistoric Africa—and surely it was—then evolution would have selected the genes of those who were best at it.

Archaeologists Bruce Bradley and James Steele heads of the Learning to Be Human Project, believe we owe it all to stone tools. Using experimental archaeology (learning by doing), Bradley and Steele hope to use before-and-after MRI brain scans of test subjects making stone tools to show that this activity may have sparked and fueled the enlargement of our brains and the then development of higher thinking and language. Perhaps when early hominins improved their lives 2.5 million years ago by making and using simple stone choppers, called Oldowan tools, they launched a brain boom. They made manual dexterity, motor control, and the impulse to modify what nature offered advantageous traits.

A million years on, *Homo erectus* were making the more sophisticated Acheulean hand axe. Producing this more complicated stone tool required more planning and hierarchical thinking. Then, by 600,000 years ago, *Homo heidelbergenesis* were making spears, cleavers, and axes of a quality and complexity that Bradley believes placed an evolutionary premium on language, visual imagination, and symbolism. Tools, he reasons, it was all about the tools and what traits the construction of them made valuable in us. "I'm willing to bet," Bradley told *New Scientist* magazine in a 2014 interview, "there wouldn't be consciousness on this planet if we didn't have flakable rocks."[20]

The Benefits of a Good Night's Sleep

Strange as it seems, where we slept and the quality of that sleep might have had a lot to with our extraordinary brain development over the last couple of million years. Australopithecines, our smaller, more chimplike predecessors, probably slept in trees, which means that they had to be light sleepers in order to keep themselves from rolling off branches and falling at night. A fall could break a bone or make one vulnerable to predators. But their descendants, *Homo erectus*, were likely too tall and too heavy to sleep safely anywhere but on the ground. Their use of fire to keep predators away might have helped make this a viable option. Sleeping on the ground and not having to be concerned with falling off a branch would have given them the luxury of deep sleep.[21]

Deep sleep, we know today, allows for longer periods in rapid-eye-movement (REM) sleep and slow-wave sleep. These phases of sleep are critical because that's when our brains are the busiest making new neural connections. It's during sleep that our brains sort and file away memories and newly acquired skills from the day before. The kind of deep sleep that allows for extensive and regular dreaming has also been tied to creativity during waking hours.[22] It's remarkable when you think about it: A good night's sleep may have been all we needed to rise up and become the intellectual giants of our planet? Let's just hope that dolphins never work out a way to get in eight solid hours of sleep per night. If so, they may end up ruling us.[23]

So which one was it? Which one of these factors took us from being smart primates to astonishingly smart primates within such a relatively short period? No one knows. It could be, of course, that it was a combination of these or entirely different factors that sparked the human brain explosion. Or, perhaps it was caused by something we have not yet imagined.

One thing seems almost certain: it didn't have to happen like this. We all might pause once or twice in our lives to reflect and appreciate that we are able to, well, reflect and appreciate things. While other complex biological features have evolved repeatedly, a brain as big and powerful as ours has come along but once. "The very large brain that humans have, plus the things that go along with it—language, art, science—seemed to have evolved only once," says evolutionary biologist Richard Dawkins. "The eye, by contrast, independently evolved forty times. So, if you were to 'replay' evolution, the eye would almost certainly appear again, whereas the big brain probably wouldn't."[24]

ARE YOU MORE EVOLVED THAN A WORM?

All of this talk of evolution and the magnificent human brain requires clarification and a disclaimer. Sure, it's very good at the things we do. But let's keep in mind that it's not very good at the things we don't do. Evolutionary success is not measured in Van Gough paintings and Shakespearian sonnets. It is about quality and consistency of performance—whatever performance survival demands. If a species works well enough in the current environment, then it is a success and it gets to live another day. If it does not, extinction is the verdict. This is why some restraint and a bit of humility are appropriate when one describes the wonders of the human brain. It is all too easy to go overboard with the self-admiration. Yes, we have the most powerful analytic and creative brains ever seen on Earth, and we have accomplished many spectacular things with them. But this does not mean that we are the undisputed champions of evolution.

Humans may be capable of scoring better than fish and worms on human-designed IQ tests, but fish and worms are here with us right now, which makes them every bit the evolutionary success we are. In the game of life, the best we can claim is a tie. Evolution is about the status of being alive or extinct *now*. It is easy to be confused about this. Many people view evolution as an ascending path or ladder that leads to continually better outcomes over time. We have bigger, smarter brains, therefore we must be the "most evolved," right? No, we are not. You are no "more evolved" than a bacterium or a mushroom.

Evolution is not synonymous with progress, because progress implies that something is getting better. Better for what? If "better" a century ago or a million years ago turns out to be ineffective today, which happens all the time to numerous species, then "more evolved" can be a death sentence. Extinction has been the most common fate by far for life-forms over the last few billion years because the environment and evolution do not consult with one another and strive for sensible upgrades and general improvement. Evolution is about environmental pressure applied to current life. Life is then carved and hammered by these blind pressures over generations. Evolution does not think up ways to improve and prepare a species for the future, because evolution doesn't think and it certainly doesn't know the future. Think of evolution as merely the indifferent result of life's constant interactions with previous environments. It's like the wind and rain that shaped some canyon wall fifty million years ago. It might seem beautiful or even intelligently shaped to human eyes, but there was no purpose or goal. What we see today in everything from plants to people to microbes is the result of what their ancestors experienced hundreds, thousands, millions, and billions of years ago.

There is no winner's podium to evolve up to, because we never know what will constitute winning tomorrow. For example, many people today think of human intelligence as a wonderful triumph of evolution. But this is nonsense. It certainly helped keep us alive so far and I'll choose my human brain over any other, but big brains were just one successful adaptation out of countless trillions that Earth's life has experienced. We may place supreme value on it today, but aren't we a bit biased as the sole possessors of such a brain? And the game is not over.

In the evolutionary perspective, there is no triumph because the struggle never ends. What if, for example, our relatively high intelligence leads directly to our extinction? Nuclear missiles we designed and built may destroy us all one day. Our dense and dirty civilization may poison and obliterate the natural world to the point where Earth's ecosystems can no longer support us. What if some of our brightest minds build an artificial intelligence that quickly surpasses our mental abilities and then promptly decides to exterminate us? If any of those things happened, then our big brains wouldn't have been such a great thing in retrospect. We would have been better off holding steady at *Australopithecus afarensis*.

NEW PROBLEMS, NEW SOLUTIONS

A fascinating aspect of the human brain is that its weakness is its greatest strength. Because of the unplanned and haphazard evolutionary history behind it, our brain is not like the key or the combination to a safe. It's a safecracker. It has not evolved to react only in specific ways when confronted with specific challenges. It evolved to think and create, to improvise and adapt and continually find new solutions to new problems.

Consider how our children are clueless and helpless for a longer period than any other species. We do not come into the world with a brain that has been preloaded with the necessary information for survival. But what it does have is the flexibility to learn and create. This has enhanced our survivability and enabled us to do amazing things. This is why it is so important for us to work at becoming good thinkers. To drift through existence without understanding the brain or developing critical-thinking skills is to live an undervalued life. Why not maximize our membership in the only deep-thinking species known to exist? The human who does not embrace her brain's many powers with enthusiasm and a sense of ownership is like a dolphin that never swims or a falcon that never flies.

With the ability to think symbolically, plan for the future, and learn complex skills, we flourished. That this profoundly creative-thinking organ came about through a winding, fragile, impossible-to-repeat sequence of make-do modifications and on-the-fly redesigns may be difficult for some to accept. But it's worth restating that the modern human brain we see today clearly is a product of just this. As I pointed out previously, it is not a miracle of perfect or ideal form and design. Far from it. But it can still manage to generate novel solutions out of thin air because it learns, collaborates with other brains, and imagines without limit. In their book, *The Brain*, biologist Rob DeSalle and anthropologist Ian Tattersall write:

> As the product of a long evolutionary process with many zigs and zags, involving ancestors who were living in circumstances that were hugely different from our own, the human brain is an affront to the principles that any self-respecting engineer would follow. And it is in this that its secret lies. For to a large degree, it is the very fact that its history has not optimized it for anything that accounts for the human brain's being the hugely creative and simultaneously both

logical and irrational organ that it is. . . . Our brains and the occasionally bizarre behaviors they produce are the product of a lengthy and untidy history: a history that has accidently resulted in a splendidly eccentric and creative product.[25]

As stated earlier, the best evidence currently points to the first anatomically modern humans emerging only about 200,000 years ago, an incredibly slim slice of existence on a 4.5-billion-year-old planet that has hosted life for more than three billion years. The entire primate order stretches back only some fifty-five million years into the past.[26] As a good thinker, you will want to keep all of this in mind. Try to remember that your recent prehistoric ancestors were very much like you. Only their environments and experiences were significantly different. Although there appears to be some surprising shrinkage over the last 20,000 years,[27] the brains we have today appear to be virtually the same model that people were walking around with in Africa more than 100,000 years ago. This fact alone should stir questions, if not concerns.

How can we expect to cope well in the high-tech, heavily populated, concrete maze of twenty-first-century life with a brain that evolved within the heads of isolated bands of prehistoric hunter-gatherers? So much is different now. Most of us don't have to worry about being eaten by large predators or spend our days scouring the wilderness for calories. We mostly sit and stare at phones, computer screens, and televisions. How has our ancestor's brain held up so well in such a radically different environment from what we knew for 99.999 percent of our existence? One answer to that is that we are *not* coping so well these days. Mental-health problems, poverty, drug addiction, crime, wars, terrorism, food-related diseases, and death by stress are all common features of the contemporary landscape. So maybe our prehistoric brain doesn't work as well as we think it does here in this modern world of our making. Perhaps our cities and current cultural behaviors are not entirely good for us.[28] But we are thriving at surviving, at least according to population figures, so the other answer has to be that we are fortunate to have a brain that is so flexible that it makes a pretty good showing in just about any time or place.

Just as our ancestors used their brains to invent the Acheulian hand axe and fire-starting skills, we used the same brain to invent the computer and develop spacefaring skills. It only seems that we are so distant from them in mental function and capability because of how we have utilized the accumulation of cultural knowledge over time. Let's be honest, it was not your brain or mine that produced spoken and written language. You are able to read this book and talk with friends thanks to the brains of your ancestors who did develop those things. I believe it can help to think of ourselves as prehistoric people deposited in the future. We may move around in comfortable cars and planes, but our brains and bodies still feel and work best when we stand, walk, and run. We may enjoy eating many "foods" that are little more than delivery systems for sugar and salt, but our brains and bodies still do best when fueled with vegetables, meat, and fruit. And it doesn't matter where we live or what we do for a living, we still rely on prehistoric brains to make decisions and cope in the world around us. Dean Buonomano, a neurobiologist at UCLA's Brain Research Institute, believes our prehistoric brains present us with a clear challenge:

> Although we currently inhabit a time and place we were not programmed to live in, the set of instructions written down in our DNA on how to build a brain are the same as they were 100,000 years ago. Which raises the question, to what extent is the neural operating system established by evolution well tuned for the digital, predator-free, sugar-abundant, special-effects-filled, antibiotic-laden, media-saturated, densely populated world we have managed to build for ourselves? . . . Our feeble numerical skills and distorted sense of time contribute to our propensity to make ill-advised personal financial decisions, and to poor health and environmental policies. Our innate propensity to fear those different from us clouds our judgment and influences not only who we vote for but whether we go to war. Our seemingly inherent predisposition to engage in supernatural beliefs often overrides the more rationally inclined parts of the brain, sometimes with tragic results.[29]

These are valid concerns, and the wise reaction to them is endless education and aggressive awareness. Become a lifelong student of the evolution, structure, and operation of your brain. The more you understand, the more rational and safe you are likely to be. Yes, we may be burdened with a brain that has been caught out of its time in modern culture. But there is nothing better than the human brain at finding solutions to important challenges, even if the challenge is to understand that brain and use it well.

Chapter 3

EXPLORING YOUR BRAIN

"Nothing is too wonderful to be true, if it be consistent with the laws of nature . . ."
—Michael Faraday, lab journal entry, 1849

F irst rule of Brain Club: You do not talk about how complicated the brain is and how much scientists haven't yet figured out. There is nothing to gain from feeling intimidated or overwhelmed by the seemingly endless list of brain components or its enduring mysteries. So make up your mind not to shrink before this challenge. Unless brain surgeon happens to be your chosen career path, it is not necessary to know every detail of every structure. Nor do we need to lose sleep over the mind-melting challenge of explaining consciousness to the satisfaction of all. Fear not, we will make it out of this chapter alive. There will be only some twenty or so names to remember in the coming pages, much less than the number of celebrity names you have committed to long-term memory, I'm sure. I understand how brain anatomy confuses people. Some of the names are weird. Books, websites, and documentaries are not always clear and consistent in the way they present the brain. *Is that "fore-brain" or "frontal lobe"? "Cerebrum" or "neocortex"? "Brain stem" or "medulla oblongata"? And what is all this lizard brain–mammal brain stuff?*

Second rule of Brain Club: If you are new to Brain Club, you have to fight any urge to skip pages in this chapter. You are about to embark on a painless, confusion-free tour of the brain designed to help you understand and forever appreciate the most important part of you. I have given lectures about the brain to young students and remember how much they appreciated a clear path to comprehension. With that in mind, I will keep the medical-school jargon to a minimum and focus only on the most important highlights, the

75

least you need to be aware of. It's crucial to know where the neo-cortex is and what it does, for example, but I think most people can be forgiven for not being fully up to speed on oligodendrocytes, microglia, and astrocytes. And, keep in mind as we move forward, good thinking demands a basic understanding of the brain. This is *practical* knowledge. You can't do your best good thinking if you have never visited the place where good thinking happens.

We have to prepare for our journey into the brain by first shrinking ourselves. Let's drop down to about the weight and width of a small flea. Yes, I know, physics wouldn't allow us to walk and function well at that size, given our anatomy. That's why we will be wearing miniature mechanical exoskeletons and environmental suits, of course. And, yes, it's bound to be dark inside a skull, which is why we have tiny flashlights attached to our utility belts. This is one of the many things the human brain does well. According to our desires, it can conjure up irrational positions or outright impos-sibilities and then explain and defend them endlessly. You might want to remember this the next time you find yourself drowning in an argument about politics or religion. By the way, to make this trip more personal and meaningful for you, we are going to imagine that it's *your* brain we are exploring. Don't worry, we'll take only memo-ries and leave only footprints. Here are the four areas we will visit:

- Brain Stem (the body-brain connection)
- Limbic System (the feeling brain)
- Cerebellum (the motion brain)
- Cerebrum (the thinking brain)

BRAIN STEM

Don't turn your head or make any sudden movements. At this moment, extremely tiny versions of you and I are scaling the ver-tical heights of your spinal cord. Your brain is now our Everest. I'll be Tenzing Norgay and you can be Edmund Hillary. Yes, it's a bit weird that you would be less than a tenth of an inch tall and exploring the insides of your own body. And I'm still not sure how we are going to squeeze through the tight confines of your brain, but we will figure it out as we go along. For your sake we'll try not to dig in too deep with our crampons and ice axes. From inside your backbone, we work our way up through the cervical vertebrae and, with some effort, manage to transit an opening at the base of the skull called the *foramen magnum*. From here, finally, we can make our attempt on the brain stem and get this expedition going.

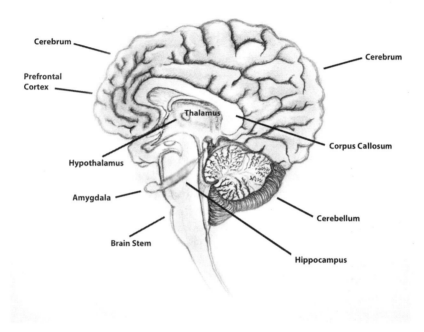

Fig. 3.1. View of the human brain cut in half, with the front of the brain at left. *Illustration by Sheree Harrison.*

The brain stem looks a bit like the stalk of some kind of vegetable. Far above us is a visible lump of tissue at the bottom of the brain. The brain stem may not be as fancy and flashy as the neocortex that we will see later, but don't let appearances fool you. This structure is profoundly important to you. It's the crucial junction between spinal cord and brain. This is the original information superhighway. It's your own Grand Central Station, with information running back and forth at all times. Messages race through here, from all points throughout the body to various regions and structures of the brain and vice versa. Whether you are playing the violin or brushing your teeth, the electrical signals that make it happen pass through here.

That lump at the top of your brain stem, by the way, is called the *thalamus*. Before we discuss it, however, let's agree not to get too caught up in assigning every brain part to its "proper" place and role. While organizing the brain into components and sections or regions can help with comprehension, we have to avoid giving too much weight to the idea of well-defined borders between brain parts. It's like looking at the Earth and seeing only two hundred or so separate nations, at the expense of the reality of a single planet intimately connected across land, ocean, and atmosphere. The brain is an organ built on connections. It cooperates and shares the workload out of necessity and for efficiency. The source of many abilities—memory, for example—is not found only in one specific region. Duties often overlap or are taken care of by multiple structures. Okay, back to the thalamus. It works like a router by relaying messages to whichever part or parts of your brain need to know. Visualize an operator sitting at one of those old-time phone switchboards and frantically plugging and unplugging cables into the appropriate sockets. But don't imagine yourself sitting at that switchboard, because the action here is fast and automatic. "You" aren't involved. If one of the feet on your large body steps on a nail while we're in here visiting, for example, the "ouch!" signal will flash by us at a speed of more than one hundred meters per second. Then, if the thalamus does its job, your foot will be pulling away from the nail, your face will grimace, and you will scream "Ouch!" all before your conscious self fully realizes what is going on. It's a great service. Thank your thalamus daily, I say. If we had to wait for the much slower conscious comprehension and reaction to immediate dangers and injuries, we would get hurt more severely and more often. But wait, there's more.

The thalamus may not write songs or solve algebra problems, but it does other things well—such as *keeping you alive*. This is the workhorse behind the regulation of many of your most basic life functions. Even if you are knocked unconscious, it will keep doing its job for you. "You" would not be able to tell yourself to keep breathing, for example, but your thalamus will.

The thalamus keeps a close watch on breathing, blood pressure, hunger, and heartbeat. It also has a say in salivation, the release of tears, and even the forming of facial expressions. Impulses to sneeze, cough, and vomit come from here, too. The latter duty eliminates the need for you to carry on lengthy inner debates with your conscious self about whether or not to puke up that warm bacterial cesspool you mistook earlier for a nice bowl of soup. Thanks to the thalamus/brain stem, the order goes out and the poison comes up. This structure also plays a prominent role in keeping you asleep or awake, as needed. And how exactly would you wake yourself up if you were asleep? Imagine how annoying and burdensome it would be if you had to remember to take a breath of air every couple of seconds. How could we ever concentrate and get work done? How many times per day might you forget to breathe and pass out? What if you had to consciously think about your heartbeat and will yourself to flex the cardiac muscle tissue in your chest every second or so in order to keep the blood flowing through your body? Forget to do that constant chore and you are dead. I'd imagine it would be quite the challenge while sleeping. Thanks to the brain stem's constant diligence, however, you rarely have to worry about these things.

LIMBIC SYSTEM

Tighten your chinstrap and gut up. We have summited the brain stem, which can only mean that we have arrived at the limbic system. No place for the meek, this is where rage and terror live. This collection of structures lies between the brain stem and cerebrum and is responsible for your various states of arousal. If someone threatens to attack you, it's the limbic system that sounds the alarm and gets your body revved up to fight or run. This is where those terrible tingles of fear and anxiety come from when you watch a scary movie or take a call from your mother-in-law. When a sexually attractive person walks by you on the sidewalk, and you walk into a light pole as a result, blame the limbic system for your shame. This is also where emotions are tied to memories. If the cerebrum is the "thinking brain," then the limbic system must be the "feeling brain." Let's explore two must-see destinations in the limbic system:

- Hippocampus (spatial awareness and short-term memory)
- Amygdala (Fear central)

Hippocampus

The hippocampus is a small, curved structure about three centimeters (1 inch) wide. You have two of them in parallel, one for each hemisphere. The name comes from the Greek words *hippo* and *campus*. Early brain researchers thought it resembled a seahorse, so they called it "hippocampus." *Hippo* for "horse" and *campus* for "sea monster." As we stand before one of yours and look over how it is placed within your brain, it's hard to get the whole seahorse thing. What is not difficult to understand, however, is the importance of this tiny organ. Without a working hippocampus, we would struggle to find the hippocampus or much of anything else and, even if we did find it, we probably wouldn't remember finding it.

This is because the hippocampus plays a big role in spatial awareness—making directional sense of surroundings. It also is responsible for laying down many of our long-term memories. This means finding your way home every day, as well as remembering what you did with the previous years of your life, depend on your two little hippocampi doing their job.

Figuring out exactly how the hippocampus goes about processing experiences into memories, and then sorting and storing them, remains a challenge for researchers. But it's constantly at work, somehow sorting and filing the moments of life, in order to serve you as best it can.

If they're deemed worthy, some short-term memories are designated to be long-term memories and are sent somewhere off into the cortex to await retrieval. Given the importance of memory to our ability to function and to the quality of life, having a healthy hippocampus is obviously crucial. Here are some things you should know about your two hippocampi. Size is the first indicator of condition. The bigger, the better is the general rule. A small or shrinking hippocampus has been associated with memory problems and Alzheimer's disease.[1] Fortunately, we do not have to passively accept the genetic hand fate dealt us. We can help keep our hippocampi healthy and functional by increasing their size. According to recent research, this can be achieved by consistently stimulating the mind and engaging in physical activity throughout life. Regardless of your age or current health status, get busy now. If you are alive, it's not too late. Read books, create art, play music, learn new skills, walk, run, swim, cycle. Start yesterday. You also want to do everything you can to avoid obesity and diabetes. If it's too late for prevention, don't give up. Attack those conditions positively and aggressively with a doctor's guidance because they are known to cause significant shrinking of the hippocampus.[2]

The Amygdala

Next we come to the nearby amygdala. You have two of these, one for each hemisphere, just like the hippocampus. Anyone can get a sense of where the amygdalae are by mentally following an imaginary horizontal line running straight into the brain from the backside of each eye to a point about four inches or so. That's where your amygdalae are nestled. Standing before one of yours, we can see that it is about the size and shape of an almond. Nothing particularly interesting stands out about it. Don't be fooled. It would be a mistake to underestimate the importance of this little bundle of neurons.

The amygdalae play a central role in the most intense, horrifying, and exciting moments of your life. An amygdala's size varies widely among individuals (1–3 centimeters)[3] based on factors such as gender and even the number of one's social connections.[4] Size also seems critical to a variety of important mental-health concerns. In 2013, researchers at the Stanford School of Medicine found that larger amygdala sizes predicted higher levels of anxiety in children.[5] A 2014 study of men found that amygdalae with a volume that falls somewhere on the lower end of the spectrum were linked to "severe and persistent aggression and the development of psychopathic personality."[6] On a happier note, another 2014 study scanned the amygdalae of people who had donated kidneys to strangers and found them to be larger and more active than control subjects.[7] These people were also better at identifying fearful expressions on the faces of others, which suggests a high degree of empathy. Think of the amygdala as two little spots in your brain with big implications.

The basic function of your amygdalae is to receive input from your senses and then, if deemed appropriate, to signal your body to react with intense aggression and/or fear. When the amygdala interprets something to be a danger to you, it immediately initiates pupil dilation, a rise in heart rate, and an increase in metabolism.

It also puts your senses on high alert and max ability. These are all things that make you better able to fight, hide, or run away.

The amygdalae can also interpret input from the senses as something very good for you and react by triggering a flood of pleasurable feelings. On top of all this, the amygdalae bring memory into the picture by "remembering" some stimuli and filing them away to enable rapid responses to them should they come along again in the future. This is why a bad experience in a swimming pool as a child, for example, might haunt the adult many years later by bringing forth feelings of anxiety or even extreme terror if she or he wades into a pool.

The amygdalae's involvement with memory and emotion means you may not only react with base fear or ecstasy, but also may be left with deep feelings of anger, sadness, or happiness tied to the event. This means memories of it will likely be more intense and meaningful when you recall them.

Perhaps the most interesting thing about the amygdala's work inside your brain is that it does its job without consulting "you," the conscious and reasoning part of your brain. When the amygdalae do their thing, "you" are once again reduced to the role of bystander as your brain and body spring to action. To save time, and perhaps save your life, the amygdala, working with the thalamus and brain stem, takes it upon itself to make you freeze in place or lurch back should you see something that resembles a snake or a poisonous spider. At the moment of the first sensory input, the need for speed trumps the luxury of further observation and deep analysis of the situation. Without this, it's highly unlikely we would be here as a species today. Can you imagine Australopithecines pondering how best to react to poisonous snakes?

A recent stroll with my family through a haunted-house attraction at Universal Studios in Los Angeles provided a beautiful demonstration of my amygdalac at work. Even though I consciously knew that real axe-wielding maniacs and actual dangerous monsters were not waiting for me around each turn, my brain's fear-response system wasn't taking any chances. Several times, an actor in scary makeup lunged at me from out of the shadows, and nearly every time, my muscles tightened, I lunged backward, and my arms shot out in front of me to block the attacker. I felt the familiar adrenaline surge that comes with being terrified. But then my thinking brain would remind me to calm down because this was all for fun and nobody was going to dismember me and my family. My heart rate slowed down, and my body relaxed—until the next actor leapt at me. It didn't matter that I was well aware that this was a safe and controlled environment. Amygdalae don't play games and don't take days off. When they receive input from the senses that translates to potential injury or death, they freak out first and leave you to ask questions later. We usually have no choice but to go along for the ride.

Another experience provided me a front-row seat to a display of the amygdala's power to not only control physiology in the moment but to also burn lasting, intense memories into the brain. Several years ago, I lived in the Cayman Islands and went snorkeling on the eastern side of Grand Cayman. This coastline is more rugged and less populated than the more developed and tourism-oriented western side. Maybe it was the soft early-morning sun, the lonely silence, and the flat seas that lulled me into a state of heightened relaxation and joy. Whatever it was, I felt extraordinarily alive that day. I was at home in the sea, soaring and slicing through the warm water with ease. I saw rays of light bounce off the shell of a passing baby sea turtle. A huge lobster watched me from its coral bunker. Funny thing, I don't recall being scared at the first sight of the shark. There was definitely no panic, no compulsion to flee. I was only curious and happy to have spotted it. At about thirty meters away, it seemed like a safe distance. I had encountered sharks without incident before while scuba diving. I knew that people are not on the typical shark's menu and that attacks are extremely rare in the Cayman Islands. I was so calm, in fact, that I gently flicked my fins to move *toward* the shark. Far from scared, my only urge was to commune with this magnificent lion of the sea.

Once I was about fifteen meters away from the shark, I began to feel different about the situation. The now-imposing size and intimidating muscular build of this two-meter (seven-foot) bull shark had made it less interesting and more intimidating. Two meters may not seem like much on land, but in the water and up close, it felt like being in a bathtub with a whale. I saw that its huge teeth were in disarray, not at all like the orderly rows one typically sees in books and on TV. Many of them protruded straight out of its mouth at a perpendicular angle. I do not know if it was the shark's overall size or those chaotic teeth, but I decided to retreat. I slowly backed away while keeping my eyes on it. It casually turned away from me and moved toward the open sea with a few slow tail flicks until it faded from sight. *Great*, I thought, *that was by far the best shark encounter I have ever had*. Little did I know that I was about to have an even better one.

For the next twenty minutes or so, I explored the waters just inside the reef where the depth was about two to three meters. I was still alone in the area except for two men in a small boat who were line fishing and spearfishing. They were a hundred meters or so down from my position. I continued to enjoy the outing. I peered into rock and coral crevices, hoping to spot an octopus or moray eel. I didn't, but I did see several colorful cuttlefish swimming by in tight formation, which was nice. And then my amygdalae took over. My shark had returned. And it was close, very close.

Before "I" even realized what was happening, my body froze in place; my heart pounded furiously; a jolt of electrocution by adrenaline shot through me; my pupils opened so wide, I could see tiny individual spikes on the shark's skin; and my muscles tensed so hard, I might have pulled something. I don't remember these specific things happening, of course, but I do remember "freaking out," and that is the lay term for these standard involuntary reactions to intense fear. The shark was less than a few meters (ten feet) away. *Gigantic* at that distance and way too close for comfort.

Fortunately, I calmed myself enough to focus my thoughts. Swimming away from it was not an option. I was being swayed back and forth like a rag doll by passing swells while the shark was virtually motionless. Funny thing, even in a state of near panic, I remember admiring its sleek form. It was hydrodynamic perfection, an aquatic fighter jet. There was no way I could swim for shore and survive if it had bad intentions. The first plan of action my brain produced in anticipation of an attack was to grab the shark in a bear hug and hold on. If I'm holding on to the side of it, it can't bite me, right? It took me about a second to dismiss this ludicrous idea. The next plan was better. I would move over to a nearby coral head and put it between me and the shark. I knew from my experience as a runner that simply facing aggressive dogs prevents them from biting you in virtually every encounter. Most predators prefer biting their victims from behind because it's safer for them. Supposedly it works the same way with sharks. So I continued facing it while moving toward the coral head. Once there, I grabbed onto the small chunk of convenient coral and hoped the shark would lose interest. It didn't.

The apex predator slowly moved toward me. This is when things got weird. I moved, it moved. We ended up playing "Ring around the Rosie" for a few laps around the coral head. It was both bizarre and terrifying. At one point, I popped my head up and saw the men in the boat. I screamed for them, then resumed the cat-and-mouse antics. I feared it was going to attack at any moment. Fortunately, it didn't. The boat came, and I put one foot on the coral head to lunge up and in the boat—not so easy with fins on. I banged my head and shin on the boat but couldn't have cared less. I was relieved to be alive. But my crash course on the amygdala was not yet finished.

About a year later, I was at an aquarium attraction in the United States with friends, and while standing before a huge tank filled with numerous species, I suddenly began to physically shake, and my heart pounded in my chest. I didn't know why. After a moment of confusion, I realized what was happening to me. A large bull shark had slowly cruised into view right in front of me. There was no doubt about the species. Bull sharks have a distinctive shape and thickness that is easy to identify. I was on dry land and safe from any harm—yet my body was in full fight-or-flight mode. Unknown to me, that morning encounter in the Cayman Islands had left me traumatized. My amygdalae had tagged the memory of the incident as special and sent it off to be stored somewhere in my brain so that the mere hint of a similar threat set off my body's alarms. It didn't matter that I was on the other side of thick glass. Better safe than sorry is the amygdala's philosophy.

The amygdala plays an important role with many phobias, too. A fascinating 2014 report in *Neurocase* tells the story of a forty-four-year-old businessman who began having seizures that damaged his left amygdala. Doctors removed it, only to discover later that they had extracted a phobia along with it. The man's previously recorded arachnophobia vanished. Not only was he no longer afraid at the mere sight of a spider, but after surgery he said he found them interesting and had no reluctance to touch or hold them.[8]

An even more unusual case is that of the woman who feels no fear. Called "SM" to protect her privacy, researchers have studied her with fascination for years because she literally is fearless. She has Urbach-Wieth, a rare disease which caused calcium deposits in her brain, some of which destroyed her amygdalae. One researcher described her condition as if someone had effectively scooped out both of them. The results of this unusual condition have been astonishing. SM says not only that she feels no fear, but also that she struggles to even imagine what fear is. She registers no alarm or terror, for example, when exposed to dangerous and poisonous animals. She even tells about a disturbing incident in which a man held a knife to her throat in a park. She says she did not panic or feel the slightest bit of fear, even as she felt the blade against her flesh. She told the man, "go ahead and cut me. I'll be coming back and I'll hunt your ass." He let her go uninjured.[9] One can imagine SM's case attracting the attention of some political or military dreamers. What might a small team of assassins or a vast army of soldiers who are incapable of feeling fear be capable of?

The amygdala is an invaluable component of the brain that has served our species well for a long time. As mentioned, without the near-instant response to potential threats that the amygdala initiates, we never would have made it this far. Keep in mind that the standard life experience of most primate family members most of the time—people included—has been to nervously and habitually look over one's shoulder in the hopes of not spotting an approaching predator or in-species rival from the next valley over. It may be comforting to imagine our ancestors as master killers, kings of the jungle, and all of that, but the reality is that we have logged far more time being the hunted than we have being the hunter.[10] And that legacy is in us.

Remember that the brain is not a bunch of outposts in isolation. Fear is not isolated to the amygdalae. This is a complex emotion that touches multiple spots. What the good thinker needs to be aware of is that the amygdalae and all subconscious fear processing present us with a downside to consider here in the modern world. The unthinking nature of reactive and instinctual fear can make us afraid of things that don't really threaten us and, unfortunately, there is never a shortage of people ready to exploit our fears for their gain. Even though we may be sitting safely in our living rooms, for example, we still can be terrorized by images of violence and disasters that are far away and have little or no chance of harming us. Video news is a master of this kind of fear exploitation. For-profit cable-news companies, for example, know that scaring their viewers keeps them watching. So, no surprise, they emphasize scary topics and events. Terrorism, murders, plane crashes, and potential disease pandemics are products for sale. Misery has been monetized. And we are the customers. It is outrageous how cable news so often distorts our view of the modern world. I encourage people to travel whenever possible in order to enrich their lives and become more enlightened and connected human beings. But a common objection I hear when I talk about visiting faraway lands is that it's just too dangerous. News media have created generations of paranoid people who think they will be murdered ten minutes after leaving their home country—which in many cases is a more dangerous place than the locales they might visit. Sure, we are somewhat of a mess with all our self-inflicted problems, but it's still not as bad out there as we are led to believe. Fight back against all those images of disease, explosions, and AK-47s in your head. Do some research. Crunch the numbers. Study the world map. There is a lot of peace and quiet being waged around the globe, too, certainly much more than the death and destruction your subconscious fears keep reminding you about.

I conducted a memorable interview in 2012 with a woman who was convinced the world was going to end soon. Why? Because a religious leader she followed had told her so. Her worldview of global civilization on fire and imploding didn't hurt, either. One only had to watch the news, she explained to me, to see that the world is almost destroyed already. Her delusion was airtight. The preacher's claim was not based on logic and evidence; therefore it was immune to both logic and evidence. Because she had never learned about the human brain's weakness for authority figures and compelling stories, and its susceptibility to confirmation bias and other cognitive landmines, she had been a victim in waiting. Her lack of critical-thinking skills and failure to learn about irrational fears made her weak and vulnerable. She was already on the hook. The preacher she trusted had only to reel her in. I remember worrying about her a lot when her doomsday date came and went. I still worry about people who fall so deep down the well of bad thinking.

It happens to all of us, though. I don't know about you, but I sometimes catch myself becoming a little too concerned about things that I know are statistically too silly to worry much about. I sit down on a plane. *Wonder if there are any terrorists on board.* I walk into a movie theater. *Hope nobody slips in through the exit and shoots the place up.* I go for a hike or a run alone in the mountains. *It would suck if I bump into a family of cannibals up here.* Even though my thinking brain chimes in and tells me instead to focus on things that are far more likely to impact my health and safety—such as what I eat for lunch every day, driving safely, and how many hours of sleep I get each night—I still have these other, much less realistic dangers in mind in part because of media fearmongering. But it doesn't mean I have to surrender to them and alter my life unnecessarily. We can override the irrational fears that haunt us. Maybe not every one and not all the time, but we can push back against many of them.

Politicians around the world exploit fears to near perfection these days. It's nothing new, of course, but it does seem that television, radio and the Internet have raised the art to an all-time high. Sometimes I wonder if campaigns in America have become nothing more than fearmongering competitions. Almost any issue can be exaggerated and presented as a potential disaster that will ruin lives and destroy the country. Voters who are unaware of how subconscious terror can well up from within in reaction to the manufactured threats delivered in carefully crafted scenes and phrases in political ads are less likely to be rational citizens at the polling station. It is simply too easy for politicians and their allies/enablers to steer us emotionally by saturating television screens with images of exploding buildings, body bags, and warnings of imminent economic, social, and moral collapse. Real issues worthy of our attention are hijacked and made into launch pads for runaway anxieties and easy political manipulation. Resist. You don't have to be cannon fodder in someone else's war on reality. Maintain a healthy level of awareness and suspicion so that it's not so easy for unscrupulous people to yank your amygdalae.

Good thinking means constantly assessing and reconsidering fears and feelings of anxiety. Think it through. *Is my fear irrational? What exactly are the reasons for me to be scared? How afraid should I be? What are the chances of this terrible thing happening to me? Is my reaction appropriate or excessive? Is the source of information about what is scaring me competent and trustworthy?* Ask these questions diligently throughout life, and you will make fewer poor decisions in the name of fear.

CEREBELLUM

Our next stop is the cerebellum, so let's head toward the back of the skull. It's at the bottom rear of the overall brain. It is small compared with the larger cerebrum all around us. About the size of a plum, it's split into two hemispheres and stands out due to its distinctive deep and symmetrical grooves running mostly from side to side. *Cerebellum* is Latin for "little brain," by the way. I suppose it's a pretty good name because it does look like a smaller brain that has been plugged halfway into the back of your larger brain. Then again, "little brain" might be misleading because the cerebellum seems to have a hand in a lot of activity. It also may have played a prominent role in the brain's evolution.

Your cerebellum's tissue is folded over even more than the wrinkled and much-folded cerebral cortex up top that everyone is so familiar with from photographs and images of the brain. The cerebellum is so convoluted, in fact, that if it were spread out, it would cover a surprising 1,128 square centimeters.[11] That's about the space two standard-size sheets of paper would take up. The cerebellum also has significantly more neurons than the rest of the brain combined, as much as 70 percent of the brain's total! And the way those neurons are wired together has not changed in more than four hundred million years of vertebrate evolution. This means your cerebellum's neurons are organized just like a rabbit's or a lizard's.[12] Unfortunately, scientists don't yet know why the "little brain" would have so many more neurons than that vaunted cerebral cortex that we keep on a pedestal for all the heavy thinking it does. This is another one of the mysteries about the brain that has yet to be solved.

Some scientists point to the cerebellum's rapid growth during the period in our evolutionary history when human mental capabilities greatly expanded and suggest that the cerebellum's role in "technical intelligence" was behind our ascension as intellectual masters of the planet. Evolutionary biologists Robert A. Barton and Chris Vinditti believe that the potentially crucial role of the cerebellum in prehistory has gone largely unrecognized and understudied. In a study published in 2014, they wrote that "technical intelligence was likely to have been at least as important as social intelligence in human cognitive evolution. Given the role of the cerebellum in sensory-motor control and in learning complex action sequences, cerebellar specialization is likely to have underpinned the evolution of humans' advanced technological capacities, which in turn may have been a preadaptation for language."[13]

Likewise, neurophysiologist Jim Bower views the enigmatic cerebellum as deserving of more attention. "It is a strange and disturbing structure," he says, "with relatively few people working on it. If you remove the cerebellum in young humans—or animals—[it] turns out there is very little noticeable effect a few months later, [which is] rather odd given how many neurons are there and how much energy the cerebellum consumes. But this [mystery] mostly speaks to how primitive our understanding of human behavior is."[14]

The cerebellum is believed to take on primary responsibility for your posture and balance, as well as any complex or fine physical movements you may wish to execute. Perhaps you enjoy playing the guitar or piano. If so, thank this impressive ball of neurons before us, because it probably has more to do with your ability than anything else. Traditionally the cerebellum has been described as the motor-control center of the brain. When a baseball player catches a ball over his shoulder while running, a circus performer juggles five bowling pins while balancing on a unicycle, or a young girl climbs a tree in her backyard, it is the cerebellum that is coordinating the effort. It sends and receives real-time information from various points around your body in order to make fast and often subtle adjustments that help you to succeed at whatever physical feat you may be attempting. Think of the cerebellum as having a ship's bridge somewhere within it. The helmsperson receives information from deck hands, observers, or computer sensors, and then makes appropriate changes in speed, attitude, and direction in order to avoid hitting icebergs or asteroids, depending on what kind of ship we are imagining. This fast back-and-forth communication with other parts of your body is crucial for everything from walking over uneven ground without stumbling or rubbing your eye without poking it out to clearing ten high hurdles while sprinting at full speed.

The cerebellum doesn't operate in isolation, however. Remember, the brain is a complex collaboration of parts. Most processes involve multiple components working as one. It also gets involved in some way with many other routine activities. Brain-imaging studies have discovered, for example, that the cerebellum lights up in conjunction with normal, everyday hearing, smelling, pain perception, and even thirst.[15] It is also involved with some memory processes. This is typical, of course. The more we learn, the more interesting the brain is.

My youngest daughter currently plays on an elite team for one of the top-ranked volleyball clubs in the United States. She is a defensive player, which means that her primary responsibility is to receive and pass balls that have been served or hit by attacking opponents. When I watch her in practice and in games, I marvel at how she is able to get to, control, and accurately pass extremely fast-moving balls to teammates. I also know that her cerebellum is ablaze with activity during every play.

The moment a ball is struck on the other side of the net, my daughter's eyes take in the image of the moving ball. Her optic nerves fire off a flash of information to her brain, which then rapidly estimates the ball's speed and predicts its trajectory. If her brain decides that the incoming ball will travel into her zone of responsibility on the court, a complex series of high-speed movements are set into motion. Her leg muscles flex, popping her feet off the court and launching her toward the precise spot where her brain has determined the ball is going to be. This part can happen so fast that it's sometimes difficult to visually track the footwork involved. At the same time this is happening, her arms begin coming together and moving into position to meet the ball at a precise point. Her forearms have to be at just the exact angle in order to accurately aim the pass to a teammate near the net, called the setter. If any one part of my daughter's maneuver is slightly off, by even a fraction of a second or a couple of centimeters, the pass is likely to be too difficult for the setter to play well, or the ball may even end up in some spectator's lap in the stands. It's amazing that her brain can process so much in so little time: The placement of her feet, the position of her hips, the degree of rotation of her shoulders, the angle of her arm's contact with the ball, the "give" she allows from her legs and arms upon impact to absorb just the right degree of the ball's power.

When she faces powerful hitters and servers, the speed and precision of these coordinated efforts can be astonishing. To her, it's just volleyball and another moment in the life of a fourteen-year-old girl taking care of business. For me, however, watching it all happen from ten meters away, it's a spectacular demonstration of the cerebellum's command over difficult activities.

Command might be stretching it, though, according to some scientists. Some question if the cerebellum is really "controlling" motor activities at all. Surely it plays an important role, but is it wrong to think of it as the command center of fine motor movements? Maybe so. While I have no doubt that my daughter's cerebellum is hard at work when she's playing her favorite game, perhaps it is "only" coordinating all those complex internal communications that allow her to play so well. Either way, I'm impressed.

Adding yet more mystery to the cerebellum, it is thought to play a prominent role not only in playing musical instruments but also with music appreciation in general. Yes, your cerebellum is a music critic. As revealed by brain scans, it reacts to the music you like and gives the cold shoulder to music you don't like.[16] It's as if your brain dances apart from your body. The cerebellum also seems to have something to do with precision hearing. Researchers have found, for example, that people with a degenerated cerebellum have difficulty discriminating between similar-sounding words.[17] Odder still, as Bower's previous quote described, people who have had their *entire* cerebellum removed due to medical need have been able to recover most of their motor skills over time. This is a primary reason that Bower and his colleague Lawrence M. Parsons, also a neuroscientist, suspect that the cerebellum plays an important but *supportive* role for the brain. They suggest in an article published in *Scientific American* that the cerebellum "is not responsible for any particular overt behavior or psychological process. Rather it functions as a support structure for the rest of the brain. That support involves monitoring incoming sensory data and making continuous, very fine adjustments in how that information is acquired—the objective being to assure the highest possible quality of sensory input."[18]

Bower and Parsons also argue that detecting activity in a particular area of the brain is not necessarily reason enough to conclude that the area is directly responsible for a specific behavior or psychological process. They write,

> most of the machinery under the hood of a car, by analogy, is there to support the function of the engine. One could generate all kinds of hypotheses about the role of the radiator in propulsion—by correlating increased temperatures to miles per hour, for instance, or by observing that the car ceases to run if its radiator is removed. But the radiator is not the engine. . . . If the cerebellum is a primary support

structure, then it does not contribute directly to motor coordination, memory, perception, attention, spatial reasoning, or any of the many other functions recently proposed. Although this theory is one of several competing to account for the new and surprising data about the cerebellum, it is clear that how we think about this brain structure—and therefore how we conceive of the brain as a whole—is about to change.[19]

So here we are. The cerebellum does a lot of . . . something. Is it mostly leading actions or supporting them? No surprise—more questions for future research to answer. On to your cerebrum!

THE CEREBRUM

It's time to leave the cerebellum and climb up to the massive cerebrum that dominates your brain. If we had to point to one thing inside your skull that makes you human, this is it. It's the big show, the real "thinking cap," and center stage for the highest levels of thought. This is the part of the brain most familiar to everyone. When we look at any standard picture or model of a human brain, it's the cerebrum that dominates the image with its distinctive, wrinkled surface.

In the cerebrum's lower portion we find the basil ganglia. This bundle of specialized cells works with the cerebellum and other parts of the cerebrum to coordinate voluntary movement. So, when your amygdalae make you tense up and lurch back the instant you spot a black widow spider on your hotel pillow, it's the basal ganglia that coordinates your conscious decision to walk to the front desk and request another room.

As we survey the cerebrum, we see that it is split down the middle into left and right hemispheres. Like the cerebellum, the cerebrum's hemispheres control opposite sides of the body. Scientists are not sure exactly why. The answer lies somewhere far back in our evolutionary history. Up close, the divide is so prominent that it almost looks like you have two brains jammed together. Don't make too much of the "right-brain, left-brain" hype, but do know that the left hemisphere is where most language processing takes place, and the right deals with making connections between bits of information and spatial awareness. The two halves communicate with each other via nerve bundles, the largest of which is called the *corpus callosum.*

Neocortex

After making our way up to the very top of the cerebrum, we find ourselves on familiar turf. This surface upper portion is called the *cerebral cortex* or *neocortex.* Before we traverse one of the valleys around us, we need to pause and take this in for a moment. It's important to appreciate what goes on here. This pink tissue that tops the human brain is the source of all great ideas, creative fantasies, piercing analysis, problem solving, language, and art. This is it—the most powerful component of the most powerful thinking machine in the known universe—and this one is yours! Not to take too much away from all of the other critical components below us, but this really is the star of the show. The source of its great intellectual power comes from the *billions* of neurons (brain cells) and their *trillions* of connections. No computer built to date comes close to the neocortex's ability to learn and to create. While I stand by previous statements about the crucial importance of all the brain's many components and the need to view the brain as an integrated whole, I readily admit that top billing for the neocortex is well deserved. Only mammals have it, and no other mammal species has one as impressive as ours. This thin layer of organic smarts is what makes us human. It took us around the world, built the towers of culture, and allowed us to finally reign over the predators that once stalked and ate us. Without this, you would be little more than a tall iguana.

I promised not to burden you with too many names, but I feel you should be aware of the lobes that the neocortex is sectioned into, so I will mention them in brief. Each hemisphere of the neocortex is divided into four areas called *lobes,* and each lobe has responsibilities for certain functions. They are:

- Frontal lobe: engages in problem solving, decision making, planning.
- Parietal lobe: receives and processes sensory information.
- Temporal lobe: plays a role in emotions, memory, language, and hearing.

- Occipital lobe: is the primary center for processing input from the eyes.

As we inspect your neocortex up close, the first thing that jumps out at us are the convolutions, the coils and twisted surface. What do all these curves and ripples do? It's certainly quite different from other animals. Mice and other mammals, for example, have much smoother cerebrums. That smoothness equates to less surface area, which means fewer neurons for them. Fish, by the way, don't even have a cerebrum, which is probably why they don't play chess and build space stations.

These wrinkles all over your brain were evolution's answer to the challenge of how to fit more surface area—to accommodate more neurons—into the limited space of a human head. In proportion to our bodies, we already have gigantic heads compared with the rest of the animal kingdom. And our craniums could not get any larger without demolishing the hip structure of our dear mothers during birth. Convoluted tissue—think of balling up a large sheet of paper so that it can fit into a small container—was the ideal solution, at least for a living system that can change only on the fly. Most interesting of all, however, is that the neurons in the neocortex—the nerve cells that create your thoughts and memories—work by establishing connections among themselves and then communicating with each other. So your neocortex's primary activity is talking to itself.[20]

Let's focus our attention up front, on the area just behind your forehead. This is the famed *prefrontal cortex*, the area of your brain that enables you to make plans, engage in so-called moral reasoning, and resist the constant lure of immediate gratification in order to pursue longer-term goals. If your brain had a CEO, this is where her office would be. Jordan Grafman, chief of the Cognitive Neuroscience Laboratory at the Rehabilitation Institute of Chicago, describes the prefrontal cortex as "a crowning achievement of the human brain."[21]

We can breach the surface here and explore within. Don't worry, we'll replace our divots. Punching through is not difficult. Your neocortex is less than half a centimeter thick (about two-tenths of an inch). Pretty thin, but don't be fooled. There is still a lot of it. So much, in fact, that it accounts for about *three-quarters* of the brain's total weight. If it were spread out on a flat surface, your neocortex would cover an estimated 1.6 square meters (17.2 square feet).[22] That's a little less than the surface area of a twin bed and a bit more than a loveseat, if that helps to picture it. Okay, it's time to see the show. Let's go inside your neocortex and seek out the real source of . . . you.

Neurons

In order to experience the magic that occurs in this region of the brain, we need to shrink ourselves even further and then dive down into your neocortex. Here we will investigate the two basic types of brain cells: neurons and glia.

The figure that is widely cited in books, articles, and documentaries for the average number of neurons in a human brain is one hundred billion. This is a huge, impressive number that seems to confirm our intellectual prowess. Slugs, for comparison, have only about 20,000 or so neurons in their whole body. How sad. Just because something is widely cited, however, doesn't necessarily mean that it's accurate. I looked into it and discovered that this common estimate is impossible to source with confidence. It turns out that the real number may be significantly lower. A study published in the *Journal of Comparative Neurology* in 2009 found that adult male brains have, on average, eighty-six billion neurons, pretty far off the popular estimate. None of the brains looked at in this study had one hundred billion or more neurons.[23]

In case you are curious as to how these researchers counted eighty-six billion neurons, they certainly didn't do it one by one; that would have taken them several centuries. Their method was to liquefy a brain in a mixer to evenly distribute its neurons and then count the amount within in a small, measured portion of the brain fluid. This enabled them to calculate a total for the whole brain. But don't worry, there is no chance of us losing our status as possessor of the planet's master brain—we still have a considerable lead on slugs. Whether one hundred billion or eighty-six billion, it's a staggering number of neurons. All together, the combined surface area of that many neurons has been estimated to equal more than three football fields![24]

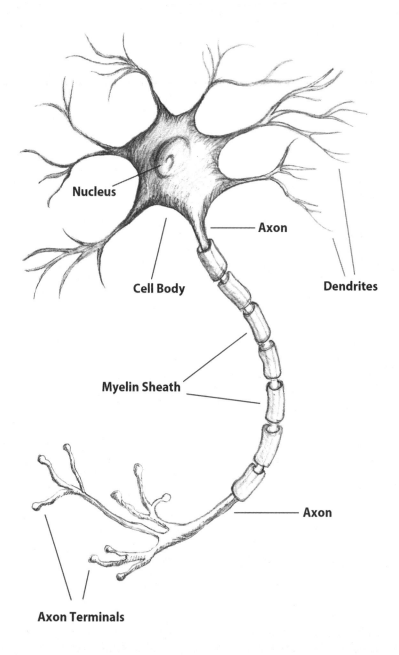

Fig. 3.2. Your brain contains many tens of billions of neurons. Their trillions of connections make all your thoughts, memories, and dreams possible. *Illustration by Sheree Harrison.*

Giving Glial Cells Their Due

We have to pay some attention to the glial cells all around us. Neurons may get most of the attention, but these cells matter, too. There is disagreement about how many are in a typical human brain. Traditionally they have been thought to outnumber neurons 10:1. But, like the popular neuron count, that doesn't seem to be backed up by good evidence. Some researchers believe the glial-neuron ratio may be closer to even.[25] Regardless, what matters is that the various kinds of glial cells serve as a crucial support system. Think of glial cells as the working-class pit crew for the superstar neurons. They are responsible for less glamorous but vital chores within the brain. Glia help neurons by nudging them along to help make connections, and keeping them in place when they do. They also supply neurons with the building blocks they need to operate, and they dispose of dead neurons. Glial cells clear debris or waste products from the brain and even defend it against foreign germs that somehow manage to penetrate the ringed security system of bone, membranes, and fluids that surrounds the brain. Beyond all that, however, new research has found that glia detect and control the activity of neurons and may even have a role in the formation of memories. Additionally, glial cells have a prominent role in many brain injuries and diseases and have been implicated in some mental disorders.

The emerging awareness of all this prompted neuroscientist R. Douglas Field, head of the Neurocytology and Physiology Unit at the National Institute of Child Health and Human Development, to criticize the $4.5 billion BRAIN Initiative, a massive effort announced by the White House in 2013 to map the entire brain. Think of it as a Human Genome Project/Moonshot for the brain. Field is concerned because the project seems to have neglected glia from the onset:

> That the word "glia" was not uttered in any of the announcements of the BRAIN Initiative, nor written anywhere in the "white papers" published in 2012 and 2013 in prominent journals outlining the ambitious plan, speaks volumes about the need for the community of neuroscientists behind the initiative to expand its thinking. . . . Neuroscientists have known for several decades that glia cause certain diseases. Nearly all cancers originating in the brain derive from glia—

which, unlike mature neurons, undergo cell division. . . . Many neuro-logical disorders that were once considered to be exclusively neuronal in nature are now also known to involve glia, including Rett syndrome, a neurodevelopmental condition that includes autism-like symptoms, the motor-neuron disease amyotrophic lateral sclerosis, Alzheimer's disease and chronic pain. The same is true for various other develop-mental and psychiatric conditions such as schizophrenia, depression and obsessive-compulsive disorder.[26]

"In any major mapping expedition," Fields continues, "the first priority should be to survey the uncharted regions. Our under-standing of one half of the brain lags a century behind our knowl-edge of neurons. I believe that answers to questions about the brain and public support for a large-scale study are more likely to come from expanding the search into this unknown territory. . . . This exploration into the 'other brain' must be done together with the proposed studies of neurons. It cannot be achieved as a by-product of them."[27]

Fields makes a compelling point. It reminds me of a key lesson learned in the twentieth century, that it is necessary to focus con-servation efforts on entire natural habitats, where possible, rather than attempt to study and save only individual species here and there. The web of life is too complex, the components too intertwined to approach any other way. This is a good way to think of the brain as well, like a vast ecosystem that will never be understood unless all of it is purposefully investigated.

Now that we have given glial cells their due, let's return our attention to the neuron that is the carrier and possessor of your thoughts and memories. As we shrink even further, the neurons near us grow larger. Don't be surprised if your amygdalae stir, because at this scale the neurons have taken on the look of exotic and dangerous monsters. Although much like any other cell—with the usual nucleus, DNA, mitochondrion, vacuoles, ribosomes, and cytoplasm—their structure brings to mind a colossal squid from the abyss or maybe a menacing, tentacled extraterrestrial from some 1950s sci-fi movie. They create endless and connected mountain ranges of strange life all around us. Finally the numbers begin to sink in. "Tens of billions" of these things is still a difficult amount to appreciate, but now, here among them, we begin to sense their col-lective presence. We see them in every direction. Above, below, and on all sides, they sparkle with electrical energy. We are astronauts now, adrift in a living galaxy. Only in this galaxy, the stars are close enough to touch. At this point, we are both a bit uneasy because this is such a peculiar place. But there is no denying the beauty of it all. Sebastian Seung, a professor of computational neuroscience and physics at MIT, describes the collective of neurons within a single brain as a poet might:

> No road, no trail can penetrate this forest. The long and delicate branches of its trees lie everywhere, choking space with their exuberant growth. No sunbeam can fly a path torturous enough to navigate the narrow spaces between these tangled branches. All the trees of this dark forest grew from 100 billion seeds planted together. And, all in one day, every tree is destined to die. This forest is majestic, but also comic and even tragic. It is all of these things. Indeed, sometimes I think it is everything. Every novel and every symphony, every cruel murder and every act of mercy, every love affair and every quarrel, every joke and every sorrow—all these things comes from the forest.[28]

Who says hard-science people can't have soft hearts?

Monstrous though they may seem to us up close, these extraordinary cells inside your brain are *you*. This is the true core of humanity in extreme close-up. These neurons achieve the unlikely and make impossible thoughts a matter of routine. By forming vast and complex networks within the neocortex, they routinely give birth to ideas that are new to the universe. There is no end to the creative potential found in here right now. No end. We have discovered the fountains of infinity within the walls of your skull. This ultra-complex and thick thatchwork of connected neurons all around us reveal your brain's source of greatness.

Best of all, these networks are not fixed. When we learn, create, and experience new things, the brain rewires itself, builds new networks. Don't pass over this too quickly. Let it sink in: you finish each day with a new brain, one that has been changed by the experiences of your previous hours at work and play. This is why the human brain has been able to do so much over millennia and will continue to do more.

Change, establishing new connections, is the secret of our success. With this in mind, let's take a closer look at how the neurons around us are linked together. One draws close to another,

which reaches for a third, and so on. Any single neuron is likely connected to many thousands of other neurons. It is easy to see that
the total number of connections for your entire brain must be huge.
Indeed, it numbers in the trillions and is always changing. Neurons
die. New neurons are born. Links are lost. New links are made. I
compared the scene around us to a galaxy. Well, in at least one way
your neocortex is bigger than a galaxy. There are far more connections between neurons here in your brain than there are stars in the
Milky Way galaxy. There could be as many as *100 trillion* of these
links in your brain at this moment. To add to this complexity, any
individual network of neurons can be on or off at a given moment,
which changes the overall makeup of your neural jungle profoundly
from one moment to the next. This is big, so give your neurons a
moment to process it: *Billions of neurons are creating millions of
networks via trillions of connections in your brain.* The number of
possible variations of your brain state is staggering, more than the
number of elementary particles in the known universe, according to
neuroscientist V. S. Ramachandran.[29]

As much as this place we are in represents the very core of
humanity, however, it remains alien to us. Scientists have only just
begun to learn how the brain works at this deepest level. Far more
mystery than understanding remains. But we can grab the precious
knowledge that science has thus far afforded us and run with it.

Let's focus on the neuron closest to us and figure out what does
what. It's a multipolar neuron, the most common of its kind. The
cell body, where the nucleus is located, is not the dominant feature.
What pops out at us are all those tentacles and that long tail. The
tentacles, called *dendrites*, spread out like thin tree branches. Kind
of like old telegraph cables, they receive messages in the form of
electrical impulses from other cells. The long tail is called the *axon*.
It transmits messages away from the neuron, toward other neurons.
Remember glial cells, the critical support crew? They are all around
us, too. In addition to supplying nutrients and attacking invaders—
let's hope they don't come after us!—glial cells provide material for
the myelin sheath that's wrapped around the axon. This fatty tissue
is an insulator that coats the axon (tail) of your neuron to speed up
the signals that travel within it. Sadly, some people lack this substance and suffer with multiple sclerosis as a result. By the way, for
all their importance and power, neurons are more fragile and needy

than other human cells. They will die fast without a steady supply of glucose and oxygen.

Up close, we can see something like bulbs, at the tip of the branches that flair out at the end of the axon. These are called *axon terminals*, and they lock onto receptors found along the dendrites of another neuron. The juncture where the axon terminal and dendrite come together is called a *synapse*. But the two ends don't actually touch. Let's make our way closer to get an even better look. Now we can see that there is an extremely small gap between the axon the dendrite.

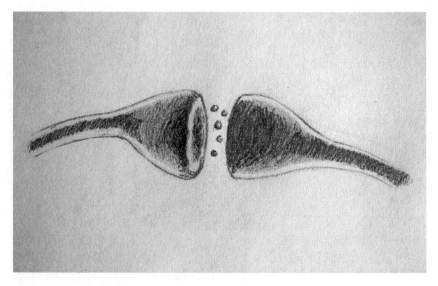

Fig. 3.3. Where the magic happens: A synapse is the point where a dendrite of one neuron comes together with an axon of another neuron for the transference of neurotransmitters. They don't actually touch but draw just close enough to share batches of chemical information. This is how thoughts, memories, and dreams are made, stored, retrieved, and connected with other thoughts, memories, and dreams. *Illustration by Sheree Harrison.*

What happens next is the strange and wonderful transfer of information that somehow produces thinking. We will wait for this particular neuron to receive and pass on a message so we can observe how it works. Brains tend to be busy places, so it shouldn't take long. Hold on, here it comes! An electrical impulse has come down from a dendrite, and now it's racing down the axon. The electrical signal reaches the end of the line and triggers the release of a batch of chemicals called *neurotransmitters*. Lean in closer. There they are! We can see these neurotransmitters, bits of information, "swimming" across the synaptic gap and into the other neuron's dendrite. Once there, this neuron has the option of sending the signal on to yet another neuron, or not. And on and on it goes throughout the brain, every moment of your life. So how do these impulses turn into complex thoughts? How do a bunch of connected neurons become consciousness? No one knows. Not yet.

Even with big unknowns remaining, there is plenty of motivation here for thinking of one's brain daily. We understand how new connections are made. We know that connections become networks, and the abundance and use of these networks make us smart, capable, and healthy. What science has discovered so far provides us with reasons to be excited. How can we not be excited? Our brains are forever changing. Your brain becomes a reflection of what you choose to do. It's the labyrinth that is forever under construction, always becoming something new, for better or worse. Pause and use your brain right now to contemplate what this means. Billions of neurons create trillions of connections, a vast system that is in constant flux. Upgrade or downgrade. It is your choice. Learn or stagnate. Explore or stay put. Grow or shrink. Your daily actions decide the size, power, and vitality of the galaxy within you. This is empowering information. I suspect that many people imagine the brain to be little more than a mushy computer that just kind of sits there in the skull and cranks out thoughts similar to the way a heart pumps blood or a kidney filters blood. Far from it. The brain is a dynamic, restless entity that reacts and grows according to what our senses feed it. It's alive with the constant flow of electrical and chemical signals. Simply being aware of this spectacle within can serve as motivation to treat it well and make as many new connections as possible throughout life. While working at home, I often glance over at the model of a human brain I keep near my workspace. When my kids aren't around to mock me, I sometimes caress it and smile. It seems weird, I know, but hear me out. When I touch the model, it helps me to "see" the real brain inside of me. I feel proud, grateful, and relieved, to be a member of the one species on Earth that has a brain capable of learning and thinking about its brain.

A GARDEN OF NEURONS

Do not be passive or fatalistic about the state of your brain. We can nurture, build, protect, and enhance it. I'm not suggesting we can all rewire our brains and suddenly start playing concerts at Carnegie Hall or collecting Nobel Prizes. But science has revealed ways in which we can improve whatever it is that mom, dad, and environment gave us between the ears. We think of our brain and its thoughts as determining what we do and what we become. But it's a two-way street. What we do also determines what our brain does and what it becomes.

"When we learn something, whether it's a French word or a salsa step, cells morph in order to encode that information; the memory physically becomes part of the brain," explains John J. Ratey, associate clinical professor of psychiatry at Harvard Medical School.

> As a theory, this idea has been around for more than a century, but only recently has it been borne out in the lab. What we now know is that the brain is flexible, or plastic in the parlance of neuroscientists— more Play-Doh than porcelain. It is an adaptable organ that can be modeled by input in much the same way as a muscle can be sculpted by lifting barbells. The more you use it, the stronger and more flexible it becomes.[30]

The first step is to take ownership of your brain and help it help you. Nurture it, grow it. I like to think of my brain as a garden of neurons. I visualize my brain cells within, and this motivates me to do things that benefit them. Just as I would water, weed, and fertilize a backyard vegetable garden, I support and protect my most important garden of all. Researchers have discovered in recent years that brain growth does not end in childhood. Not only can we make new connections between existing neurons throughout life, but we can grow new neurons in adulthood, too.[31] How can anyone hear this remarkable news and not act on it?

Do new things. Learn a new language. Help a child with her math homework. Bike through a neighborhood you have never been to. Explore a new forest. Play a new sport, even if it's just alone or with your family in the backyard. Take a shot at juggling, sculpting, or painting. Write poems and essays. Have a friendly debate with a friend. Visit a new museum. Sign up for piano lessons or simply blow made-up tunes on a cheap harmonica at home. The key is to work your brain so that it can work for you.

In his book, *The Brain Training Revolution*, neurologist Paul E. Bendheim presents five useful brain-building rules.[32] Here is an abbreviated look at them:

1. **Pace.** Start slow with new activities and make progress. Think of your brain as a muscle. Give it a chance to respond and grow stronger.
2. **No magic pills.** "Real brain training benefits come from real cognitive work," Bendheim writes. "Just as there is no magic pill to make your body without effort, there is no magic pill to train your brain. Developing it takes work."
3. **Do it all.** "Exercise all of your brain. . . . Don't be a specialist; be a Renaissance person."
4. **No day off.** One way the brain is not like a muscle is its ability to work every day. There is always something we can do. Remember, listening to music, reading a book, or playing a game are brain workouts that don't necessarily feel like work.
5. **Mix it up and find joy.** "Vary your cognitive routine, be open to new social opportunities, explore your world, and, above all, have fun. Having fun while exercising your brain is crucial."

Make sure to experience what is going on inside your skull every day. *Feel* those dendrites and axons reaching out for one another. *Hear* the crackle of all those fresh neurotransmitters flowing across newly created synapses. *See* that beautiful tangle, the dense neural jungle inside of you. Accept your responsibility for its health and growth. Most important, perhaps, seek joy and a sense of accomplishment in the process of making your brain stronger and more alive. As a human you are defined by your brain, so be sure to include its needs in your daily work, thoughts, and routines. Nurture and challenge your garden of neurons to the fullest and enjoy the satisfaction of becoming ever more human with each passing day.

Chapter 4

WHO'S MINDING THE BRAIN?

"We are what we repeatedly do. Excellence, then, is not an act but a habit."
—Will Durant, *The Story of Philosophy*

I t would be negligent to promote the appreciation and good use of a human brain without touching on its basic requirements for optimal performance, safety, and longevity. After all, the best critical thinkers in the world will make unnecessary errors in judgment and suffer avoidable intellectual problems if their brains are nutritionally abused, trapped inside sedentary bodies, or starved of novel stimulation.

The first thing everyone must know when it comes to taking care of the brain is that it is an extremely needy and greedy organ, unlike any other in this regard. The blood and oxygen demands are exceptional, if not outrageous. Based on a 150-pound person, the adult brain accounts for only about *2* percent of bodyweight but requires *20 to 25* percent of the body's blood supply. Imagine one small closet draining a quarter of the electrical power for an entire house. And the brain's supply of fresh blood must be kept constant, or severe problems arise fast. This demand ratio is even more imposing early in life. The brain of a child under the age of five requires between *40 to 85 percent* of resting metabolism (energy) for maintenance![1] This is why it is essential for babies and children to consume nutritious food in great quantities, consistently. Children who don't get enough run the risk of brain developmental problems that could burden them for life.

The body knows what matters, even if we don't. If a crisis strikes, other organs will be deprived of blood in an attempt to keep the brain, the number one priority, functioning. All of this blood flows to, through, and away from the brain via a special system of blood vessels so extensive they would stretch out more than 100,000 miles if placed end to end. The main conduits for this constant blood flow are the two carotid arteries, which are located on each side of the neck. This is why chokeholds that apply pressure to these arteries are so effective in mixed-martial-arts bouts. When the neck is caught in a leg or arm squeeze that disrupts the airway, it is relatively easy to hold your breath long enough to mount one or two escape maneuvers or maybe hold out for the bell if it is late in the round. But when the hold compresses a carotid artery, blood flow to the brain can be quickly compromised, and it's lights-out.

The reason the body places so much emphasis on getting lots of blood up to the head is because it is literally a life-and-death matter. An oxygen-starved brain dies frighteningly fast. Everyone should be aware of how vulnerable the brain is compared with other organs. The heart, for example, can be reactivated nearly twenty minutes after stopping. But brain cells can be stressed and compromised within seconds, which results in unconsciousness. Irreversible brain damage can occur within minutes, with death following soon after that if the flow of oxygen-rich blood is not resumed.[2] Every day of our lives, we are all just seconds away from ultimate disaster if something goes wrong with that blood/oxygen supply to the brain.

BRAIN FOOD

Most people associate diet and exercise with physical attractiveness, body-fat percentage, maybe heart disease or diabetes, but little else. All important considerations, no doubt, but they should not overshadow the needs of the brain. Given its importance to our lives, the brain ought to be the priority motivation for eating well and staying active. It has been well publicized for many years now that exercise reduces the risk of cancer, diabetes, heart attacks, and more. This is great, of course, but there is much work to be done to raise awareness about what diet and physical activity do for (or against) the brain's health and performance. Don't wait for popular culture to eventually tune in and figure out the science on this, because the sooner you adjust your lifestyle, the better off you'll be. Let the herd catch up to you.

Fernando Gómez-Pinilla, a professor of neurosurgery and physiological science at UCLA, has studied the intersection of food, exercise, and the brain for years, and he believes that we should all think of food as a "pharmaceutical compound that affects the brain."[3] This makes good sense. Understanding how food interacts with our brains can motivate us to make better choices at mealtime. By making smart decisions about food, we can help ourselves to think better, improve our moods and memory, perform better at school and work, and reduce the risk of Alzheimer's disease. "Diet, exercise and sleep have the potential to alter our brain health and mental function," adds Gómez-Pinilla.[4] "This raises the exciting possibility that changes in diet are a viable strategy for enhancing cognitive abilities, protecting the brain from damage and counteracting the effects of aging."

As we saw when we explored your brain in the previous chapter, the synapses that connect neurons are the key component to learning and memory. So what can we eat to help them do their job? "Omega-3 fatty acids support synaptic plasticity and seem to positively affect the expression of several molecules related to learning and memory that are found on synapses," explains Gómez-Pinilla. "Omega-3 fatty acids are essential for normal brain function. Dietary deficiency of omega-3 fatty acids in humans has been associated with increased risk of several mental disorders, including attention-deficit disorder, dyslexia, dementia, depression, bipolar disorder and schizophrenia. A deficiency of omega-3 fatty acids in rodents results in impaired learning and memory."[5]

Moreover, according to Gómez-Pinilla, children with higher amounts of omega-3 fatty acids in their diet have been shown to perform better in school, in reading, and in spelling, and they have fewer behavioral problems. Okay, sounds good. But what are omega-3 fatty acids and where can we find them? Omega-3s are polyunsaturated acids necessary for many normal and routine functions but our bodies can't produce them, so we have to get them from plants and animals. Common sources of omega-3 acids include salmon, walnuts, spinach, kale, flaxseed, and kiwi fruit. Frank Sacks, a professor at the Harvard School of Public Health, offers this advice: "We do not know whether vegetable or fish omega-3 fatty acids are equally beneficial, although both seem to be beneficial. . . . For good health, you should aim to get at least one rich source of omega-3 fatty acids in your diet every day. This could be through a serving of fatty fish, such as salmon, a tablespoon of canola or soybean oil in salad dressing or in cooking, or a handful of walnuts or ground flaxseed mixed into your morning oatmeal."[6]

THE GREEN BRAIN

In old age our brains shrink, memory declines, and cognitive abilities slip. But there is something relatively easy we all can do to push back against these losses: Eat green leafy vegetables daily. In a remarkable study published in 2015, scientists tracked the eating habits and mental abilities of 954 older women and men for two to ten years. The researchers adjusted for gender, education, age, smoking, genetic risk for Alzheimer's disease, and even physical activity in order to zero in on the impact of diet alone. The results should encourage us to run to the nearest farmer's market and buy vegetables.

Participants who ate more green leafy vegetables had less cognitive decline. And they didn't have to consume tons of the stuff. Those who ate just one to two servings each day "had the cognitive ability of a person *eleven years younger* than those who consumed none."[7] The specific nutrients in the food that scientists suspected were helping the brains were vitamin K, lutein, folate, and beta-carotene.

It is crucial to make the connection between your food and your brain. Our daily decisions about what we eat have a direct and lasting impact on the brain's health, productivity, and longevity. When it comes to diet, perfection is not necessary, so don't use your inability to be perfect as an excuse not to do better. Don't think in terms of success or failure, just follow the science and do your best. Forget the fad diet of the month and give up the dream of finding nutrition in a pill. Simply strive to make smart choices about most meals. As a general rule, avoid foods your brain hates and seek out those it loves. Get this right about 90 percent of the time, and you will be doing well.

I suggest finding or creating recipes that are good for your brain and your personal tastes. Eating is one of life's pleasures; don't turn it into a burden. You just have to accept a basic underlying truth:

a sharp, focused, healthy brain with staying power for a lifetime requires a diet dominated by vegetables, fruits, and quality protein. It is not going to do as well if it habitually is abused with too much added sugar and too many calories from the wrong kinds of foods and drinks. Remember, your brain is trapped in your body. It can't leave in protest. It's up to you to maintain your body for it and for overall health. A brain that is perched atop a body in crisis is compromised and has little chance of doing its best work.

Think before you set foot inside a grocery store. Remind yourself of the importance of what you are about to do. Much of your brain's health and ability to perform literally depend on decisions you make while walking the aisles. Buy things that are good for your brain, taste good to you, and can be inserted into your daily routines with relative ease. For example, I have my own personal recipe for a blended drink that I have been making four or five times per week for several years. It's simple to make in a blender, takes me no more than a few minutes to prepare, and I love the taste. Consuming it is a treat, not a duty. I also enjoy a minor emotional reward with each swallow because I know I am consuming something good for my brain and body. This drink recipe is nothing high-tech, weird, or complicated. It's the anti-snake-oil concoction, an evidence-based, scientific brew straight out of nature. I simply toss the following ingredients into a blender: two-thirds of a cup of spinach or kale; a quarter cup of blueberries and/or blackberries; half a banana; half a carrot; a few nuts; a teaspoon of flaxseed; three or four ice cubes; and half a cup of water. To be clear, this is a *blended* drink, and not a juice with the pulp filtered out. And, as someone who strives to be a good skeptic, I am duty bound to mention that all of these ingredients are subject to change should evidence come along that shows they are not good for me. Although I like it, I certainly don't *believe* in this recipe with any sort of zealotry or unwavering loyalty. I don't view it as the great panacea humanity has been searching for. I don't believe it will boost my IQ or singlehandedly keep my memory sharp forever. I don't go around tapping strangers on the shoulder and insisting they try it. I simply think it is a convenient and enjoyable drink that makes sense for me given what I currently understand about the brain and its nutritional needs. Give it a try.

I also make my own custom bags of unsalted trail mix to take with me on the go so I can nibble and graze and help keep my brain going strong throughout the day. Typically I choose cashews, peanuts, walnuts, almonds, and dried cranberries. This convenient snack suits my tastes and prevents me from getting so hunger-crazed that I end up gorging on some great-tasting but brain-crip-pling industrial sludge that has been engineered to exploit my prehistoric desires for sugar and salt. Bananas work for me as well. I think of them as one of nature's perfect mobile snacks. Find what-ever works for you within the bounds of sensible, science-based nutritional advice and go with it. Your brain and body are likely to thank you for it in the long run by not rebelling against you or slowing down too early in life.

THE LURE OF FAKE FOOD

One can buy highly questionable pills and powders that promise to do everything science-based food and drink choices do—and more. The current supplements market is ready and waiting to sell virtu-ally anything to anyone for every need. These high-priced products might be more convenient and save some time but are unlikely to do the job nearly as well as real food, if at all. Some might even hurt you. We will explore more about this in chapter 10.

Keep in mind that our brains and bodies evolved over millions of years to use real plants and animals as fuels and building mate-rials. This is why many experts recommend obtaining omega-3 acids from actual fish or nuts rather than pills, for example.[8] There were no health-supplement shops in Africa a million years ago. Our ancestors ate actual food that they had hunted and gathered, and we have the bodies that evolved accordingly. This is not a blanket condemnation of all supplements in every situation. If needed and used correctly, they can be invaluable to some people, of course. But for most of us, most of the time, real food is the safe and sensible foundation for brain and body nutrition.

Bittersweet

It has become evident in recent years that added sugar in foods and drinks presents a significant health risk not only to our bodies but also to our brains. Not all sugar is bad, of course. Within reason, the complex sugars/carbohydrates found in fruits, for example, are a fine energy source. The brain can't think and conduct its other business well without a steady supply of energy from food, including sugars/carbohydrates. It is the consumption of mountains of simple sugars found in junk food that is the primary problem. High in calories and low in nutritional substance, they race through the body, providing a brief spasm of metabolic excitement and confusion. This is followed by a long lull and an excess of leftover calories. Consuming an abundance of added sugar on a regular basis is terrible for the body and the brain, like trying to run a Ferrari on saltwater and lighter fluid.

A 2013 study published in the journal *Neurology* indicates that a diet chronically high in added sugar, even in people who are currently healthy and nondiabetic, exerts "a negative influence on cognition [thinking], possibly mediated by structural changes in learning-relevant brain areas."[9]

Did you catch that? "Changes in learning-relevant areas." I don't know about you, but I don't want cheesecake and sodas changing any learning-relevant areas in my brain. There is also reason to be concerned about falling into a trap of sugar addiction.[10] No, I am not presenting sugar as the new *Reefer Madness*–style bogey man, but there is some disturbing information we all should be aware of. For example, consuming high-sugar foods and drinks on a regular basis can activate the release of opioids and dopamine, which can lead to cravings and a cycle of consumption that may be significantly harmful to the body and the brain. On the other hand, foods that provide a substantial hit of natural glucose do just the opposite. A decent dose of glucose from real food, the preferred fuel of the body and the brain, usually leaves us feeling satisfied and full. But a meal that is heavy with sucrose, the common added sugar that does nothing good for your body or brain, can actually trigger intense cravings for more of it.[11]

It might help us to keep in mind sugar-consumption recommendations from the World Health Organization (WHO) and the American Heart Association (AHA). WHO's official stand for years has been that no more than 10 percent of an individual's daily calories should come from added sugar. In 2014, however, it announced that a consumption rate of 5 percent of daily calories was ideal.[12] For an adult with normal body mass, 5 percent of daily calories would be approximately six teaspoons (twenty-five grams) per day. Obviously, small children should consume significantly less than that. Additionally, the AHA says women should keep added sugar to six teaspoons or less per day, nine teaspoons per day for men.[13] Based on four grams equaling approximately one teaspoon, those figures equate to about twenty-four grams for women and thirty-six grams for men. For perspective on these recommended limits, the United States Department of Agriculture reports that the average American eats and drinks thirty teaspoons or 120 grams of sugar per day.[14] That's twenty-four more teaspoons than the WHO recommends. Try to be aware of how fast sugar consumption adds up over the course of a day. The biggest culprits tend to be sugary drinks, and many experts advise reducing or eliminating them from your diet first if cutting sugar is a goal. Such drinks go down fast, don't leave you feeling full, and can be loaded with obscene amounts of added sugar. Remember that this is not just about calories and belly fat. It's about learning and thinking, too. A 2014 study on how drinks sweetened with sugar or high-fructose corn syrup affected the brains of rats found that daily consumption of these drinks led to reduced memory and learning ability. Interestingly, the impairment was most severe on adolescent rat brains.[15]

I'm not obsessive about nutrition, but I do routinely check labels for sugar content, and you might consider doing the same. I am often stunned to discover how much added sugar is in popular foods and drinks. For example, according to its label, an eight-ounce (237 mL) can of Arizona Iced Tea with Lemon Flavor contains twenty-four grams of sugar. A common bottle of Pepsi (20 fluid ounces/591 mL) contains sixty-nine grams of sugar. A venti (8 ounces/237 mL) Java Chip Frappuccino from Starbucks comes with eighty-eight grams of sugar. And a large chocolate milkshake at McDonald's packs a whopping 120 grams of sugar! That's nearly *five times* the World Health Organization's recommended limit for an entire day in one drink.

More bad news: Researchers at the University of California's Childhood Obesity Research Center (CORC) found that many of these popular drinks have higher doses of the bad sugar, fructose, than claimed. The study looked at thirty-four popular drinks, including Coca-Cola, Pepsi, Dr. Pepper, Mountain Dew, and Sprite, and found that the fructose mixture was even worse than is generally believed. "We found what ends up being consumed in these beverages is neither natural sugar nor high-fructose corn syrup but a fructose-intense concoction that could increase one's risk for diabetes, cardiovascular disease and liver disease," said Michael Goran, director of the CORC and the study's lead author.[16] "The human body isn't designed to process this form of sugar at such high levels. Unlike glucose, which serves as fuel for the body, fructose is processed almost entirely in the liver where it is converted to fat." It's probably best to just avoid these liquid nightmares all together. Americans each now consume *forty-five gallons of sugary drinks per year*. But, as all good thinkers know, popular doesn't necessarily mean smart.

If you are a daily coffee drinker, take an inventory of what else may be drifting about in there with your coffee. Consider reducing the amount of added sugar or syrups if necessary. I often write at a Starbucks coffee shop and sometimes, in between sentences, I glance up and see scary, terrible things. Parents routinely come in with small children, as young as seven or eight years old, and buy them gigantic, syrup-injected latte monstrosities topped with whipped cream. Not the best way to start a day. Come to think of it, not the best way to start a life. While it is good that Western nations have begun to face up to their problems with soaring obesity and diabetes rates, relatively little attention is given to the specific impact of nutritionally poor diets and added sugar on the brain. This is a clear mistake, given what's at stake. Many researchers are suggesting that excessive sugar in the diet may be linked to an increased risk of depression and dementia. This is not to suggest that added sugar is the cause of all depression and dementia, of course, but it does seem increasingly clear that added sugar is probably a contributing factor in many cases.[17] Excessive added sugar also seems to reduce levels of brain-derived neurotrophic factor (BDNF), a chemical substance in our brains that is necessary for learning and memory. Researchers have found, for example, that animals with higher levels of BDNF learn faster and remember better than others.[18]

STAND UP AND MOVE

Everyone knows that physical activity is good for the body. What wasn't known until as recently as the mid-1990s, however, is just *how good* exercise is for the brain specifically. This goes far beyond trying to drop a few pounds to look good on the beach. Abandon old notions of what it means to be fit and healthy. Obesity, for example, is *less* deadly than a stagnant lifestyle. One large study found that it is better to be overweight and somewhat active than thin and sedentary. After tracking more than 300,000 European women and men for several years, researchers discovered that a lack of physical activity was associated with twice the number of deaths than those caused by obesity. Even walking at a fast pace for as little as twenty minutes a day has a significant positive impact on lifespan, the researchers concluded in a paper published in 2015.[19]

It is clear that one of the many benefits of physical activity is living, and, let's be honest, your brain won't be of much use to you in a dead body. So if you need more motivation to get moving, here it is. Another remarkable study, also published in 2015, looked at the activity patterns and life spans of more than half a million mostly middle-aged people.[20] The researchers found that people who did not exercise at all were at the highest risk of early death—no surprise—but those who did just a little, significantly less than the recommended 150 minutes[21] of moderately intense exercise per week, reduced their chances of premature death by 20 percent.[22] People who hit that recommended 150-minute mark had a 31 percent less chance of dying early. Want even better odds? Consistently exercising 450 minutes per week, an average of a little more than one hour per day, were *39 percent* less likely to die early than those who never exercised.[23]

In addition to aerobic activity, strength training can also benefit the brain. After all of the research I've done, I strongly encourage a mix of both moderate aerobic training and more intense workouts with weights or strenuous exercises that utilize your bodyweight, such as push-ups and pull-ups, at least three times per week (as long as your doctor has given you the go-ahead, of course). The days of thinking weight rooms are dungeons for meatheads are long gone. Resistance training is for everyone. It is necessary to stress your body with challenging movements if you want to better your chances of being strong and active late in life. A big part of maintaining a healthy brain is getting up off the couch and going outside to try new and challenging activities. But that becomes much more difficult if one becomes hobbled by weakened bones and muscles. Weight training can also lower blood pressure and improve cardiovascular health, which reduces the risk of vascular dementia and strokes.[24]

Work by neuroscientist Teresa Liu-Ambrose and her colleagues at the Brain Research Center at the University of British Columbia indicates that weight training is an invaluable compliment to aerobic activity—and that it's never too late to pick up a dumbbell. Her lab conducted the world's first randomized, controlled trial that looked at cardio exercise *and* weight training on cognitive functions in women with early dementia. "We found that only participants who did weight training showed significant improvements in both memory and executive functions," explained Liu-Ambrose.[25] "This is in contrast to earlier studies on healthy participants that showed cardio exercises to be beneficial. When we performed neuroimaging, we also observed areas of the brain responsible for memory and executive functions showing more neural activity after weight training. . . . Weight training, even as little as once or twice a week, can minimize the rate of cognitive decline."

The evidence linking physical activity to overall health and brain fitness is loud and clear now. An excellent study in Finland tracked about 1,500 people over twenty-one years and found that those who had exercised in some way at least twice a week on average were 50 percent less likely to have dementia than those who were not regularly active.[26]

The Mobile Brain

It is now clear that a moving, sweating body makes the brain happier, healthier, and more productive. But of course this only makes sense. Where is your brain? It rests inside a body that has evolved for mobility. Sitting still for hours, long periods of inactivity, the absence of physical exertion for months and years are all abominations to the human being. For the last two million years or so of our existence, we spent virtually all our waking hours upright and on the move. That's who we are. In prehistory, *Homo erectus* walked from Africa to at least *two other continents*, and then anatomically modern humans walked, ran, and paddled from Africa to *every continent* except Antarctica. They did this over many generations, of course, but they did it. Up until relatively very recently, the best if not only way to find more food or a better, safer place to live was to walk or run to it. Like it or not, you belong to a standing, walking, and running species. Not a sitting and idling species. The sedentary life goes against the grain of what it means to be a human. It's a violation of programming, a rebellion against our inherited genetic legacy—and it kills us. Researchers have discovered that sitting for prolonged periods increases our risk for heart disease, diabetes, and cancer—*regardless of regular exercise!*[27] Even if one walks, runs, swims, or cycles regularly, it may not be enough to compensate for long days of sitting. "It is not good enough to exercise for thirty minutes a day and be sedentary for twenty-three-and-a-half hours," warns David Alter, senior scientist at the Toronto Rehab, University Health Network, and Institute for Clinical Evaluative Sciences.[28] "Avoiding sedentary time and getting regular exercise are both important for improving your health and survival."

Moving is as much a part of who we are as a life-form as are language and tool use. The harsh truth is that anyone who does not engage in physical activity on a consistent basis is doing a poor job of being human. This currently popular lifestyle of sitting in chairs, cars, and couches during virtually all our waking hours is new and unnatural. Even as recent as a hundred years ago, most people were on their feet and physically active due to the demands of their daily work. The typical lifestyle today is not what we are supposed to be doing—and our brains and overall health suffer as a result. If you have desk job, adapt to this new awareness and stand up. It's not difficult to elevate a computer monitor or laptop to eye level. There are stand-up desk platforms for sale now or you can do it yourself with a stack of books or a cardboard box. Just make sure you do it. It may feel strange at first, but you will adapt in short time. After all, it's what your body and brain prefer, according to science.

QUIET THOUGHTS ON MEDITATION

We know moving helps our brains, but what about being still, very still? Although scientists have much to learn, a number of compelling studies indicate that meditation may be a worthwhile activity for our brains and bodies.[29] Some research shows, for example, that meditation may provide help to those struggling with chronic pain or depression and improve overall feelings of well-being for many others. It might even change the physical structure of the brain and slow its aging by stimulating the growth of new neurons. One study scanned the brains of twenty Buddhists who regularly practiced meditation and found that two areas of their brains had greater volume than a control group's. If future studies confirm that meditation does indeed grow more neurons, then we should all be doing it because more neurons generally equates to a healthier and longer-lasting brain. Pay attention to this area of research in the coming years. It may turn out that this ancient ritual has a valuable role in the modern world.

The way forward on the matter of physical activity is rather simple. Whenever possible, stand rather than sit; walk rather than stand; and run rather than walk. Just do what you can when you can. Regardless of what popular culture has led you to believe, it is both natural and necessary to move. Those who belong to the sedentary, nonactive herd are not living their lives as fully awakened and alive human beings. This may sound overly dramatic and maybe even a bit harsh, but it is true. Now, before any readers defensively tune me out due to fears that I am going to insist that everyone must run marathons or join a powerlifting club, there is reassuring and encouraging news about all this. According to science, *some* exercise is profoundly better than *no* exercise. Even a little activity, engaged in over months and years, can do many good things for your brain.

Neurologist Paul E. Bendheim believes it may help people to think of their brain as another muscle—the most important one of all:

> Just as you can bulk up a specific muscle through exercise, you can enlarge and strengthen specific regions of your brain through exercise. Sound crazy? Moderate physical exercise, such as walking for at least 30 minutes three or more times per week, increases blood flow to the brain, enlarges your frontal lobes, and adds new memory-recording neurons in your hippocampus. Through exercise you can replenish some of the cells lost in the aging process. Moderate aerobic physical exercise, the type that makes you breathe faster and increases your heart rate, is the most powerful trigger of new cell production in the brain.[30]

Exercise stimulates or aids in the growth of new neurons. Reflect on this for a moment. Let it soak in. *Exercise grows your brain.* It leads to a better, healthier brain, and your brain's fate is your fate. What more motivation does one need to get moving? Remember the glial cells we saw while exploring your brain? They were the ones critical to the ability of neurons to create networks and send signals. Research indicates that exercise also keeps these vital worker cells healthy and productive thanks to increased blood circulation in the brain.[31] And remember BDNF, the important chemical substance involved with learning and memory? As I mentioned previously, high-sugar diets appear to hinder the production of it—that's bad. However, exercise seems to boost BDNF levels—that's great.[32] This is mind-blowing information, and we are fortunate to live in a time that allows us to know about it and utilize it. Physical activity—in conjunction with learning, good nutrition, and consistent exposure to stimulating environments—produces more neurons and more connections between those neurons. This thickening of the brain jungle is known as *cognitive reserve*, and the more of it, the better. The greater the *synaptic density*, the more protected we are from the devastating effects of old age, strokes, and brain diseases. It's like money in the bank. The more of it you have, the more you can afford to lose before it starts to complicate your life. Therefore it makes sense to do things that force your brain to react by growing to keep up. It may well save you and your family from an abundance of misery farther down the road. Scientists have discovered during autopsies, for example, that brains of some elderly people had been ravaged by Alzheimer's disease for years—but no one noticed while they were alive! Their brains had stayed a step ahead of the symptoms because positive lifestyle choices and behaviors resulted in sufficient cognitive reserve to compensate for the destruction taking place.[33]

Commit to a life in motion: Go easy at first and make up your mind to stick with it for the long haul. I know personal fitness trainers, and they tell me that about half the people who start exercising quit after a few months or so. Don't be in the half that quits. Pace yourself and consistently reflect on what your efforts are doing for your brain as well as your body. If you find that you hate it, don't give up. Instead, try other kinds of activities, exercises, and sports until you find one or a few that you enjoy. Just be sure to stay active, and you'll reap a lifetime of rewards.

PHYSICAL ACTIVITY AND DEPRESSION

Depression is more than sadness. It is a crippling illness, a silent fog of misery that dampens and smothers the spark of life within. At its worst, depression robs its victims of the ability to feel the joy and hope that make the ride of life worthwhile. Not always easy to diagnose and awkward to talk about openly, depression wreaks havoc on both individuals and entire societies. According to the World Health Organization (WHO), more than 350 million people worldwide suffer from it at any given time.[34] In addition to the emotional devastation it wreaks on individuals and their families, there is a substantial negative economic impact on society of $30–$44 billion per year.[35] Bad as they are, such figures are almost always too low, given the reluctance of many people to speak up about their depression and seek treatment. Sadly, the world continues to lag in both awareness and care for mental illnesses, including depression, despite its prevalence. As many as two-thirds of people with depression don't even understand that they have a treatable condition. Of the people who are diagnosed, about half never received treatment.[36] A significant number of people who do reach out for professional help end up being misdiagnosed because many health professionals are not adequately trained to recognize and treat depression.[37] In some countries, more than 90 percent of depressed people receive no help.[38] And none of this considers all the collateral damage depression inflicts on the families, friends, and coworkers of depressed people. The worst outcome of depression is suicide, of course. About a million people take their lives each year, globally. That's some 3,000 people per day.[39]

A popular view of depression in the public these days is that it is simply a "chemical imbalance." While that is certainly better than blaming demonic possession or a lack of will power, the problem is far more complex than simply running a little low on dopamine, for example. Due to the staggering complexity of the brain, its chemistry, as well as each person's unique experiences and interactions with changing environments, this global pandemic of silent pain has not yet been solved by medical science. Although some medications and treatments work well for some people, there remains a great deal of mystery about the causes of depression and how best to help those who suffer with it. One thing we can be certain of at this time, however, is that physical activity helps many sufferers. The science is clear: Consistent physical activity makes people less likely to come down with depressive symptoms. And exercise often helps those who are depressed to feel better.

This is important information for everyone because some degree of depression will hit all of us at some point in life. The most unfortunate are those who struggle with severe clinical depression throughout their lives, of course, but even people who do not have that burden will feel depressive symptoms triggered by negative events such as the death of a friend or family member, job loss, or a failed romantic relationship. Fortunately even mild exercise such as walking for twenty or thirty minutes per day seems capable of both treating and preventing both minor and major depression for many people. This is not a flippant tip from a fitness enthusiast or someone's wishful thinking. It is a conclusion backed up by numerous studies produced after decades of scientific work.[40] Physical activity is a powerful way to push back against this cruel darkness and bring help to one's brain when it needs it the most.

This makes some form of physical activity a must for all who desire to get the most from their brains and feel as good as possible as often as possible while navigating through life. An important 1999 study by Duke researchers showed that a consistent exercise routine can be every bit as effective as prescribed medicine for severe depression.[41] John J. Ratey, an associate clinical professor of psychiatry at Harvard Medical School, points his patients and students to that study with enthusiasm. Ratey writes about it in his book, *Spark: The Revolutionary New Science of Exercise and the Brain*:

> The results should be taught in medical school and driven home with health insurance companies and posted on the bulletin boards of every nursing home in the country, where nearly a fifth of residents have depression. If everyone knew that exercise worked as well as Zoloft, I think we could put a real dent in the disease. The beauty of exercise is that it attacks the problem from both directions at the same time. It gets us moving, naturally, which stimulates the brain stem and gives us more energy, passion, interest, and motivation. We feel more vigorous. From above, in the prefrontal cortex, exercise shifts our self-concept by adjusting . . . serotonin, dopamine, norepinephrine, BDNF, VEGF, and so on. And, unlike many antidepressants, exercise doesn't selectively influence anything—it adjusts the chemistry of the entire brain

to restore normal signaling. It frees up the prefrontal cortex so we can remember the good things and break out of the pessimistic patterns of depression. It also serves as proof that we can take the initiative to change something. This paradigm holds true for exercise's effect on mood in general, regardless of whether we're depressed or coping with some nagging symptoms. Or even if we're just having a bad day.[42]

THE SLEEPLESS BRAIN

Sleep ought to be the easiest thing we do, right? We get tired, we lie down, and we fall asleep. What's the big deal? And since we spend a third of our lives asleep, one would think that we all would be very good at it. Unfortunately, it must be more difficult than it looks because millions of people can't seem to get it quite right. Forget credit-card debt, student debt, and the national debt. Sleep debt is the real crisis, accounting for illnesses, accidents, deaths, economic waste, and the loss of vast amounts of intellectual potential. The toll that insufficient sleep takes on the human body is clear and well documented. The risks of cancer, diabetes, heart disease, strokes, and obesity go up as the hours of nightly sleep go down.[43] Poor sleep habits have now been linked to increased risk for Alzheimer's disease, too.[44] Additionally, too little sleep can also pull down our moods. So much so, for example, that a bad night's sleep can have a greater impact on one's enjoyment of the following day than do problems with marital status or income.[45] Do not make the mistake of underestimating the impact of sleep deprivation on the brain. No matter what you are doing, a sleepy brain is not likely to be at its best.

Not only can lack of sleep make some routine activities like driving more dangerous, make our bodies less resistant to infections, and help us along to an early grave,[46] it can also negatively impact the workplace culture by eroding various social graces and moral restraints. A psychological study published in 2011 found that employees were more likely to show negative and unethical behavior at work if they had received less than six hours of sleep the previous night.[47] Previous research had found that sleep deprivation doesn't affect logical reasoning but can hinder the function of the prefrontal cortex, which helps us to regulate negative feelings and hostility toward others. It makes sense, then, that resisting negative urges in the workplace might be more difficult for sleep-deprived workers than it would be for well-rested ones.

Once again, a little education and awareness would probably go a long way. Sure, almost everyone already knows that sleep quality and quantity affect mood and physical energy. But there is so much more to it than just that. Perhaps if more people understood how important sleep is to their ability to learn, remember, create, and think, as well as how significant it is to their long-term brain health, they might make it more of a priority. Before writing this section, I asked several people to give me their off-the-cuff answer to a basic question: What is sleep? The most common response I heard was some version of, "Sleep is rest." It is understandable that people would say this, of course. It's not wrong. Sleep is a form of rest. But so much more is going on during sleep that "rest" hardly scratches the surface. Unfortunately, this information is not included in the educational experience of 99 percent of people. So they assume sleep is a form of hibernation or cryogenic suspension, just a nightly vacation for the body and brain. The reality, however, is that while we sleep, our brains remain hard at work.

Scientists have known for a long time that some parts of the brain are active during sleep. But they thought this was related only to breathing, keeping our hearts pumping, and other basic physiological maintenance duties. Then researchers figured out that sleep held some great importance with learning and memory. But what? While sleeping, our brain sorts memories, selecting some for long-term storage. It even reenacts or practices many of the things we learned or experienced during the day in a way that activates the same areas of the brain that were involved when it happened while awake. It does this in order to lay down memories deemed important so that they will be available in the future.[48] This gives weight to the old adage that it's better to get a full night's sleep after studying for a little bit than it is to have an all-night, caffeine-fueled cram session the night before a test. This is exciting stuff. Our brains are *physically* changing during sleep. There is more than electrochemical business underway. Dendrites, those little spikes we saw sprouting from neurons during our brain tour, grow and make connections during sleep toward what appears to be the formulation of long-term memories.[49] Neurologist Bendheim sums up the profound importance of sleep:

We often think of sleep as not doing anything, but . . . while you sleep, parts of your brain are hard at work building stronger, lasting memories and practicing tasks you learned the day before. Sleep improves all other critical brain functions—learning, thinking, planning and executing complex goals, creativity and problem solving. Restful, restorative sleep is essential for overall health. You will live longer and in better brain and body health if you pay attention to sleep.[50]

An important 2014 study looked at the role of sleep and false memories. Researchers found that sleep deprivation not only can impair real memories but also raises the risk of developing false memories.[51] This is no trivial finding. The impacts of memory systems derailed by lack of sleep can be large and small. Imagine being on trial for a crime you didn't commit, for example, and the prosecution's star witness stayed up all night watching pay-per-view movies in the court-provided hotel suite. Imagine arguing with a sleep-deprived spouse about something you are pretty sure you never did. But maybe you did do it and aren't sure because neither one of you got much sleep the previous night. And, finally, midnight snackers take note, scientists recently found that rats who woke and ate during their normal sleep times scored lower on memory tests.[52] Let there be no doubt about it: Bad sleep is incompatible with good thinking.

WHAT ARE THE BEST TIMES FOR GOOD THINKING?

Can chronobiology, the study of our natural twenty-four-hour cycles, help us use our brains better? It would seem so because scientists have tracked circadian rhythms and tested the mental performance of subjects with interesting results.[53] They found that daily fluctuations of *body temperature* have a significant impact on mental abilities. When our temperature dips below 98.6°F/37°C, we aren't as sharp and thoughtful. The best times to do challenging brain work seems to be mid-morning to noon and again between 4 p.m. and 10 p.m. Brains tend to bog down between 2 p.m. and 4 p.m., but not because of lunch as you might suspect. Researchers accounted for size of midday meals or skipping lunch all together and found that body temperature was the more important factor. So maybe a hot shower or a brief workout to warm us up is the answer if we need more afternoon brainpower.

A surprising twist to this is that the *downtimes* in our circadian rhythms may be the best time for creative work.[54] Perhaps when our brains are less capable of maintaining strict discipline of focus, they tend to roam more freely, which can lead to greater creativity. Therefore it may be best to try to match the type of task to the current state of our body. So when one is warm and alert, balance the checkbook. When one is sub-98.6° and a bit groggy, work on those song lyrics.

The key takeaway points of this chapter may be easy to understand and agree with—make smart food choices, exercise, and get enough sleep—but those three challenges can be difficult to consistently achieve in the long term. Modern life is demanding and complex; distractions and temptations are everywhere. The best motivation to get on track and stay there is making sure you are fully aware of the physical brain in your head and why it matters so much to treat it well. Once we accept the connection between the state of our brain and the quality of life, the easier it becomes to do the right things. Unfortunately, the brain is hidden away behind walls of bone, flesh, and blood. We can't see it in the mirror every morning. But it's always there, nevertheless, and it's being shaped every moment, for better or worse, by the decisions we make.

Chapter 5

BRINGING HUMAN VISION INTO FOCUS

"Hope clouds observation."

—Frank Herbert, *Dune*

A few years ago, I gave a talk in Southern California about one of my books, *50 Popular Beliefs That People Think Are True*, which surveys many extraordinary claims from a skeptic's perspective. I'm a fan of all things related to space, and alien talk of any kind is rarely boring, so I was pleased when UFOs came up during question time. I covered the usual bases that one does on this topic. I encouraged my audience to remember that the *U* in *UFO* is supposed to mean "unidentified," which is what most UFO sightings are. Just because we may see something in the sky that is not an obvious plane, helicopter, or hot-air balloon doesn't mean it must therefore be a spaceship from another world. *Unidentified* means unidentified, not (yet) identified. Sometimes we just have to live with unanswered questions. Yes, one can make a reasonable case for intelligent life possibly and even probably existing elsewhere in the universe, but raising possibilities is far from showing that it does or that any of them are here.

I poked a few holes in the ancient-alien-astronaut idea, touched on the real story behind the Roswell story, and offered some more-likely-to-be true explanations for alien-abduction claims. But for some people those explanations are never enough because they *saw it with their own eyes*. Personal experience, as veteran paranormal investigator Benjamin Radford stated in chapter 1, is probably the most common reason people believe in unusual and unlikely things. For this reason, I make a sincere effort to explain that we are all capable of having profound, realistic, exciting, scary, even life-changing experiences *within the confines of our skull*. These events may feel real in every way—without actually being real. It is important for people to understand the basic limitations of the human vision system. For example, it is virtually impossible to accurately gauge the size and distance of an unidentified object in the sky. It easily could be significantly closer or farther away than one thinks it is. This means that the "gigantic object" an eyewitness observed maneuvering at speeds that were much too fast to be a conventional Earth aircraft could easily have been nothing more than a child's balloon blowing around in the air much closer than the observer realized. If we look up to the sky and see the clear outline of an airplane, we can probably accurately judge if it's relatively close or far away. Why? Because we know the approximate size of airplanes and our brain will do an instant calculation based on planes we have seen before. (Although even this can easily go wrong if one happens to be looking at a flying scale model of a real aircraft.) But if it's an object with an outline we can't readily identify, size and distance errors come fast and easy. This is why so many people mistake the planet Venus for an alien spaceship in our atmosphere, despite the fact that it's more than seventy million kilometers away.[1]

We like to think that "seeing is believing," but often it is the exact opposite: *believing is seeing*. Much of what we "see" and how we make sense of the world around us is shaped by our past experiences and current beliefs. A primary tenet of good thinking is that we can't always trust our own senses, especially vision. First of all, we are set up to see things we want to see or expect to see because the brain is always looking for shortcuts to save us time and cut down on information clutter. When you look out at a typical street scene, for example, you may think you are seeing most of the people, cars, and buildings fairly well at any given moment. But you aren't. Not even close. Your brain only shows you one very small area of focus in any given moment. Whether or not you realize it, more than 99 percent of the scene is not in focus. It can't be. It's too much data for your brain to tackle at once. Your eyes dart around to various points to build a helpful though incomplete image in the brain. These eye movements are called *saccades*. In addition to this, the brain plugs in elements it assumes are there and feeds you a somewhat made-up and mostly fuzzy image of the entire scene. This happens every time we open our eyes. It tries to make the scenes we look at useful and sensible as well as provide them to us as fast as possible. Most of the time this system works well for us. Sometimes, however, it makes us miss important things or believe that we saw something that was never there.

Fig. 5.1. Most people assume our vision is similar to a camera that has the ability to take in a broad scene with most of it in sharp focus. The reality is far different, however. Our actual field of focus at any one moment is extremely limited. Outside of our casual awareness, our eyes compensate for this limitation by rapidly darting around to numerous points of focus. Our brains also fill in gaps in visual data by showing us what "should be there" or "probably is there." Compare these two versions of the same Paris street scene. The top image is a standard photograph. The bottom image suggests how the same scene would look to a human observer who is looking at the two people walking in the lower right corner. With this in mind, it is easy to understand why eyewitness accounts are unreliable. *Images by the author.*

Our vision is, to a degree, at the mercy of our beliefs and memories. Don't you think it's odd, for example, that people who already believe in UFOs, Bigfoot, angels, and ghosts are usually the same people who claim to see those things? Why don't skeptics have these encounters at the same rate as believers? A person who believes that extraterrestrials routinely visit the Earth and sees an unidentified object in the sky is more likely to "see" an alien spaceship because that person's brain might easily fill in a few details to the scene that it assumes should be there—but are not in reality. Finally, there is this clincher against relying too heavily on eyewitness accounts: We don't really *see* anything. I'll explain that in more detail shortly.

Fig. 5.2. If you see a white triangle and a white square—even though they are not actually there—then you should know better than to ever place total trust in your brain's vision system. I created the black objects in a way that suggests the existence of a triangle and a square, but nothing is there other than suggestions. It has been well established by psychologists, brain scientists, and magicians that we often see things that do not exist in the way our brain presents them to us. Sometimes we see things that do not exist at all. This is important to remember when you observe something strange or someone tries to convince you to believe in something strange that he saw. It's crucial for good thinkers to know that our eyes are a relatively small part of a theater production that is the brain's visual system. *Illustration by the author.*

It has been my experience that people are naturally fascinated about human vision and are eager to learn all about how it works, yet they still find it difficult to let go of the trust and confidence they place in what they see. Most people are quite sure that their own eyes, at least, would never lie to them. But all eyes lie. Well, not the eyes, actually. It's the brain that is less than honest. Every time we—you, I, and everyone else—open our eyes, we are being conned. The brain shows us a version or interpretation of the world we direct our eyes at. Anyone who claims to have seen something extraordinarily weird or important must address this problem. We aren't reliable that way, and the best that can be said about people who think differently is that they just don't appreciate the limitations of human vision. One may "see" a ghost, no doubt. I'm sure many sane, intelligent, and honest people have. But that doesn't mean the ghost is really there. Maybe it is, but you, me, or anyone else "seeing" it is just not a good enough reason to believe in its existence.

I have learned that it is not always easy to cast doubt upon a claim based solely on an eyewitness account, because people can be easily offended. They interpret a skeptical reaction to their eyewitness account as a personal attack, as if it is to say that they are dishonest, are stupid, or can't see well. It has nothing to do with any of that, of course. It's about having that human vision system. This is yet another reason why all of us should learn early in life just what is really going on inside our heads. The vast majority of people alive on Earth right now think that their eyes and brain serve up an honest and accurate image of the reality before them. They have no idea that this is completely wrong. Because societies fail to teach Human Vision 101 to every young child in every school, the erroneous assumption that human vision is reliable lives on with no end in sight. It is considered common sense that we see what we look at. But scientists know differently. For example, when people look at a scene, they feel that they see it in detail and would notice any significant changes. But this confidence is unwarranted. Numerous experiments have shown that significant and large-scale changes can be made to scenery and people are likely to miss it.[2]

While I was interviewed about supernatural and paranormal beliefs on a radio show, the host asked me an outstanding question. What would it take, he asked, to convince me that ghosts are real? Would seeing one in my bedroom, for example, be enough? I thought about it for a moment and then answered. "No," I said, "it wouldn't be enough." No matter how real the image of a ghost before me appeared to be, no matter how convincing it was in the moment, simply seeing it would not be enough. My first thoughts would be that I was experiencing a dream, an imagination, pareidolia, a hallucination, or a hoax. Wait, that's not completely honest, of course. My first thought after seeing a ghost in my bedroom would be to run or hide under the bed. But my second thought would be those other, more-likely explanations. I hope that I would do the sensible thing and doubt myself. I would seek confirmation from somewhere other than my lying eyes. No doubt the experience would be unsettling and I might even *feel* that it was real for the rest of my life. Intellectually, however, I would have to stop short of concluding that ghosts are floating around among us because science has revealed to us that human vision is unreliable in many ways. Even though I could not disprove the existence of ghosts or ever be completely sure it was all in my head, I would still need more evidence. This is not stubborn, closed-minded skepticism at its worst. This is skepticism at its best. When we understand and appreciate how the brain perceives the world around us, we will know better than to give over total trust to our senses when the stakes are high.

YOUR BRAIN IS A TIME MACHINE

Now that we know our brains *interpret* input from the eyes and *construct* images, we understand that we are not watching a live video feed, as most people think. It's more like we watch a reality-based play staged for our benefit by the brain. Even weirder, we see scenes that do not exist in reality at that moment. The brain shows us the future, or it tries its best to.

Have you ever wondered how professional baseball players hit fastballs traveling in excess of ninety miles per hour? You didn't really think they did it by "keeping their eye on the ball" did you? With apologies to baseball coaches everywhere, it is impossible to visually track a fast-moving ball—or spear, fist, or car, for that matter. It can take as long as a tenth of second for the brain to process visual information. This is way too slow to duck, hit, or catch a high-speed object coming at you. We "see" it only because the human brain fills in or makes up information to help us. The baseball player's brain shows him the *projected* or *expected* location of the ball even though it's not there yet. Our brains make up lost processing time by creating an image that we couldn't really have seen, because it hasn't happened yet. Without this ability to "see" a ball or projectile farther along its path than it really is, every game of catch in the backyard likely would end in injury. "The image that hits the eye and then is processed by the brain is not in sync with the real world, but the brain is clever enough to compensate for that," explains Gerrit Maus, a cognitive neuroscientist and expert on visual perception.[3] "What we perceive doesn't necessarily have that much to do with the real world, but it is what we need to know to interact with the real world."

The brain fills in a lot more than just fast-moving baseballs for us. Even when we casually scan the world around us, the brain takes many liberties. It only makes sense that our brains would do this. We can't possibly process and store every bit of information in most scenes we look at, so our brains don't even try. Our brains never evolved that capability, because doing so would overload our neurons and blow up our networks. Besides, most of what we look at is not useful information. The solution is to edit the input liberally in the name of efficiency. We don't need to visually process the tiniest pebbles on a path in front of us, just the ones big enough to potentially trip us. There is no reason to see every leaf in great detail on a hedgerow that contains thousands of leaves. We definitely don't need a high-definition mental rendering of every hair on the dog across the room. Most of the time, we just need to see that it's a dog. We do not see like a camera. We do not capture clear, focused images of wide areas before us. It is easy to overestimate what we are taking in visually, and this makes people unnecessarily susceptible to judgment errors regarding what they did or did not see. Compromised focus on the periphery is one thing, but what about your blind spot? Or didn't you know that each one of your eyes has a blind spot? It exists because of the way our eyes evolved. Remember, evolutionary changes happen on the fly. A whale cannot be placed into dry dock for upgrades, gazelles can't give up the ability to run from predators while nature tinkers with their ligaments and tendons, and vertebrate eyes couldn't go dark while the pressures of natural selection shaped them.

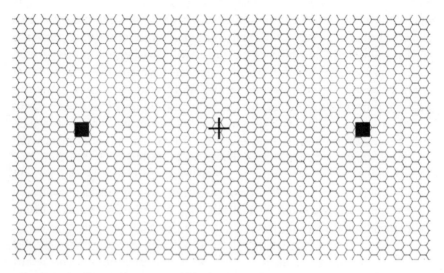

Fig. 5.3. An illustration to reveal blind spots in human vision. *Illustration by the author.*

Look at the image on this page. Now hold the page at arm's length from your face, close one eye, and focus on the crosshairs at the center of the image with your open eye. Now slowly bring the book toward your face. Keep looking directly at the crosshairs. At some point, one of the black squares will vanish. When this happens, stop and hold the book in that position. The black square has met with your eye's blind spot. It will reappear when you move the book backward or forward from that point. We all have a blind spot in each eye because the optic nerves passes through the lining of photoreceptor cells, which leaves a small area without the ability to receive light input. But that vanishing black square or the fact that you have a blind spot is not the most interesting thing going on here. The most important takeaway is that your brain is *showing you something that is not there.* Repeat the experiment. Keep looking at the center of the whole image and notice that the underlying grid pattern continues through the area where the black square was—even though there is no background grid pattern there. Do the experiment again if necessary. Pay close attention to the area where the black square should be but has vanished. What do you see? A ghost outline of the black square? A tiny black hole devouring all light and information? No, you likely see the continued grid pattern. What's going on? How can we see this grid continue over the area where the black square is? Remember that it's still there on the page, obscuring the grid pattern, whether we see it or not. This is something that your brain does all the time. It's filling in information that it is not getting from the eyes because it assumes it should be there. You brain is showing you an image that is not real but that it expects to be there. Good thinking demands that we be aware of standard brain behaviors like this because they can trip us up in unexpected ways throughout life.

We also have a tendency to miss unexpected things *even while staring right at them*. Scientists Christopher Chabris and Daniel Simons imagined and produced a brilliant experiment that exposes a common human vulnerability named *inattentional blindness*. Every good thinker needs to know about it and watch out for it. Simons and Chabris asked volunteers to count the number of passes with a basketball executed by one of two teams in a short video. While not overly difficult, the task does require some concentration because players move around while passing the basketball. When the video ended, test subjects gave their pass counts. But that was just a ruse Chabris and Simons used to get them to focus their attention. The real test was whether or not they saw the gorilla. Gorilla?

Yes, during the video, a person in a full-length gorilla costume walks into the shot, in plain view, walks between the players, pounds its chest, then exits to the left. The gorilla is on screen for about nine seconds. It may surprise you to learn that about half of the people who participate in this experiment fail to notice the gorilla, even though the ball they are watching closely passes directly in front of the stationary gorilla. No good thinker should ever forget that as humans we are capable of staring at and focusing on a scene right in front of us yet miss something as large and out of place as a gorilla.[4]

Even though our visual system routinely fails to show us everything that is there and routinely shows us things that are not there, it still works well for us most of the time in most situations. What we all need to guard against are those moments when the brain presents to its owner the image of a man in a blue shirt killing someone across the street or a twelve-foot-tall mystery primate foraging for berries in the backyard. It is in moments such as these—moments that have a strong influence on your belief whether or not something unexpected is happening—that we must remember our basic brain knowledge and ask ourselves how far we can trust this crafty and creative visual system. Maybe that murderer did have on a blue shirt, and maybe Bigfoot really was in your backyard, but can you be 100 percent sure? Vision and doubt go well together when good thinking is the goal.

Fig. 5.4. The fascinating "Checker-Shadow Illusion" illustrates how our visual system presents us with images that are not faithful to reality. Our brains make assumptions, adjustments, edits, and embellishments in an attempt to be efficient and feed us useful information. Usually it works just fine. Sometimes it causes trouble because it means we cannot always trust our eyes. The Checker-Shadow Illusion is a creation of Edward Adelson, a professor of vision science at MIT in the Department of Brain and Cognitive Sciences. The "A" and "B" squares are the same color. If you don't see it that way, it's because your brain is making assumptions based on cues in the scene, causing you to see the "A" square as significantly darker than the "B" square. It is not, however. Both squares are the same color. If you struggle to see this, even with the added bars on the right image, cut out two holes in a sheet of paper that reveals only the two squares in question and you will see that they are the same color. *Images © 1995, Edward H. Adelson.*

SEEING, HEARING, AND SMELLING . . . NOTHING

Not only do our brains routinely remove and fill in information meant to be true to our surroundings, they also occasionally present us with wholly manufactured sights, sounds, smells, and other sensations that have little or nothing to do with the reality around us. Hallucinations can fool us so well because they run through the brain in much the same way as legitimate sensory input is processed. For example, the brain of a blind person experiencing a visual hallucination may be active in the same region as a person with sight who is looking at something. It's as if the brain fools itself.

According to the US National Institutes of Health, there are many causes of hallucinations. Some include being under the influence of alcohol or drugs or suffering from dementia, fever, epilepsy, mental disorders such as schizophrenia, and even liver or kidney failure. They neglected to mention one, however. *Being human* is another common cause of hallucinations. Experts consider them to be quite normal in some situations. Seeing or hearing the voice of a recently deceased person you knew well, for example, is not uncommon and is probably associated with normal grieving.[5] Given the way our brains work when it comes to sensing and processing the outside world, we should not view hallucinations in general as unnatural or odd. Leading neuroscientist V. S. Ramachandran addresses this in his book, *The Tell-Tale Brain*: "Indeed, the line between perceiving and hallucinating is not as crisp as we like to think. In a sense, when we look at the world, we are hallucinating all the time. One could almost regard perception as the act of choosing the one hallucination that best fits the incoming data."[6]

Several years ago, I spent four days and nights alone on a small, deserted island in the Caribbean. I had imagined it as a secular vision quest or spiritual retreat minus the spirits. It was to be a break from the noise of civilization that would allow me some time to relax and think. Thanks to the constant shuffling of rats and land crabs in the bushes around me, however, there scarcely was a single quiet moment on my "deserted" island, especially at night. A few times, I thought I might have heard the distinct sound of human steps. But when I peeked out, no one was there. I suppose my brain was just working overtime to find an important pattern within the noise.

One night, while sitting on the dark beach and looking up at a spectacular display of stars, I heard someone yell my name above the sound of the waves and wind. I didn't think I heard it; I *heard* it. My head turned immediately toward the direction I believed the voice had come from. I was certain it happened, so much so that I looked around to see if I could find who had called out to me. I was even worried that drug smugglers may have come ashore until I realized that it was unlikely that they would know my name. Finally, after confirming that I was still alone on the island, I looked within. "Must have imagined it," I told myself before putting the yell out of my mind and returning to my tent. I knew enough about the human brain to conclude that it had probably been an auditory hallucination, perhaps brought on by the unique loneliness I was experiencing. I had placed myself in the highly unusual situation of going days and nights without hearing another human voice. I missed my wife and children. I missed TV, the Internet, and my bed. My brain probably was somewhat disoriented and perhaps stressed by all this, so something odd fired off within as a result. I'm still open to the possibility that someone was on the island, shouted my name that night, and then slipped away out to sea. But the more reasonable explanation is that it was all in my head.

Good thinking requires us to be aware of how our brains can and do present us with experiences that feel completely real but are not. We should also understand that such events are a normal part of being human. They are common to all societies, all time periods, and all varieties of people, so keep this possible explanation in mind when weird things happen to you or someone close to you. The good thinker remembers the limitations and the unexpected potential outputs of human sensory perceptions. After all, it does little good for one to embrace critical thinking and adopt a stance of positive skepticism, only to squander that foothold in reality by placing too much trust in uncertain sights and sounds.

Chapter 6

DON'T FORGET HOW MEMORY WORKS

**"Every act of perception, is to some degree an act of
creation, and every act of memory is to some degree an
act of imagination."**
—Gerald M. Edelman, Nobel Prize–winning biologist,
The Brain

The classic zombies of horror fiction are something more than animated corpses. They are unique monsters who frighten us in a slightly different way because they are people without memories. Zombies may move about but are not considered alive because they have no identity, no consciousness. The zombie is a creature who has been stripped of its past and now exists on pure instinct. The next meal is all that matters. Only people who have been robbed of their memories could be described as "the undead." Perhaps we shudder at the thought of zombies not only because they are decomposing and hungry but also because we sense in them something about our own vulnerability.

Memories are fragile things. Even minus the bite of a zombie, we forget so much. Our past continually recedes farther from us. Precious memories grow dimmer each day. And the loss of memory equates to losing whatever it is we think of as being truly alive. Does this mean we die a little bit with each lost memory? I think so, because our past is our life. Without it we are the undead.

Understanding how memory works and doing whatever we can do to protect it would seem to be high priorities given what is at stake. Yet once again—other than psychologists, psychiatrists, and neuroscientists—few people are ever taught the basics of human memory in school, and almost no one bothers to learn about it on her own. As with so much of the brain, there is still a great deal of mystery surrounding memory. For example, no one knows precisely how an experience is encoded and stored in our brains. Fortunately, however, more than enough is known right now to both fascinate and motivate us to make immediate lifestyle changes if needed to aid and protect our brain's memory systems.

The first thing everyone needs to know about memory is that there is nothing like an organic hard drive or DVD recorder at work, busily recording and documenting life experiences for later retrieval. Human memory is radically different from that. Misconceptions about memory are the norm, so don't feel bad if your understanding is inaccurate. What matters is that you gain a realistic conception of memory so that you can reduce the rate of memory-related errors in life. As covered during the tour of your brain in chapter 3, sensory information—a few sight or sounds, for example—comes into the brain and is routed to the hippocampus—the little seahorse in your brain. The hippocampus combines that information into a single memory, then, if it's deemed worthwhile, it is passed on to be stored somewhere in the brain as a long-term memory. If, however, the event or experience is extraordinary or dangerous, the information is also likely to pass through the amygdalae, where it will be tagged or coded as something special. This is why you are more likely to remember kissing a cute classmate or being punched by the school bully than you are some random math quiz from eighth grade.

The hippocampus also plays an interesting role in the retrieval of memories. It acts as a kind of news editor or film director that gives us a version of our past intended to be relevant and useful in the moment—accuracy be damned. The priority is to help us cope in the present and plan for the future rather than relive our past with perfect precision. It's very important that we understand this. Present and future needs trump the past in the name of immediate utility. Again, this is something a good thinker would need to know about when trying to make the best possible decisions about important matters that involve memories. It is vital to second-guess our recollections and seek corroborating evidence whenever something important depends on a specific memory.

A telling 2011 study showed just how far the public's understanding of memory strays from the scientific consensus. Researchers asked people basic questions about memory and then compared their answers with those of an expert panel made up of sixteen professors who each had more than ten years of experience in memory research.[1] Some key results:

Video Memory

- 63 percent of the public surveyed either strongly agreed or mostly agreed that human memory "works like a video camera, accurately recording the events we see and hear so that we can review and inspect them later."

vs.

- Zero percent of the experts strongly agreed or mostly agreed that memory is like a video camera. An overwhelming majority, 93.8 percent, strongly disagreed.

Confident Testimony

- 37.1 percent of the public strongly or mostly agreed that "the testimony of one confident witness should be enough evidence to convict a defendant of a crime."

vs.

- Zero percent of the experts strongly or mostly agreed with this. 93.8 percent strongly disagreed.

Hypnosis

- More than half of the public surveyed, 54.6 percent, strongly or mostly agreed that "hypnosis is useful in helping witnesses accurately recall details of crimes."

vs.

- Zero percent of the experts felt the same way about hypnosis and witnesses. A majority, 68.8 percent, strongly disagreed.

Perhaps the most important thing we all can remember about memory is that it is *associative* and *constructive*. One memory is tied to other memories, and all memories are custom-built, stitched together by the brain from portions of available information—real or imagined, related or unrelated. In short, our memories are creations that may or may not be based upon real events, and the brain's filing system is far from ideal. Although the hippocampus plays a prominent role in memory, it is not "the place" where all memories are stored. They are spread around throughout the cortex. Memories also are not stored in anything approaching the orderly system utilized by computers. That would be far superior to what we have and would save us from our struggles to remember things when we need to. In fairness, however, computers didn't have to come together over hundreds of millions of years via genetic mutations and the pressures of natural and sexual selection. Remember, living populations don't get the luxury of a pause or time-out for more sensible reengineering. They have to keep living and functioning as changes take place.

Associative memory is an often illogical system, far different from the way we would organize memories if we could design the brain. We certainly don't handle our computer data this way. Imagine if you saved your family's secret spaghetti-sauce recipe in a file labeled "May 3, 2015" because you happen to have worn a red shirt on that date and spaghetti sauce is also red. This associative filing system in our brains is why a specific smell or sound may bring up a memory even though the smell or sound wasn't an important element within the original experience. For example, the smell of diesel exhaust sometimes reminds me of Kathmandu, Nepal. There are many more relevant and important things than diesel exhaust associated with that fascinating city, of course, but my brain has encoded an association, and the smell triggers my memories of breathing there, which then brings up many more memories of my time spent there. For the same reason, it is common for certain songs to stir up memories from our lives when those songs were hits and were heard often. We remember the song and then, thanks to our brain's system that files away memories based on associations, we may remember events and people from the time we heard the song long ago.

We need to accept that our brains are not filled with isolated clusters of data neatly filed away according to date or subject to await prompt retrieval when needed. New York University psychology professor Gary Marcus describes our memory as *contextual* and believes it may have evolved as a hack, "a crude way of compensating for the fact that nature couldn't work out a proper postal-code system for accessing stored information."[2] But, he explains, our brain's way does have its virtues. "For one thing, instead of treating all memories equally as a computer might do, context-dependent memory *prioritizes*, bringing most quickly to mind things that are common, things that we've needed recently, and things that have previously been relevant in situations that are similar to our current circumstances—exactly the sort of information we tend to *need* the most."

"We're more likely to remember what we know about gardening when we are in the garden, more likely to remember what we know about cooking when we are in in the kitchen. Context, sometimes for better and sometimes for worse, is one of the most powerful cues affecting our memory."[3] According to Marcus, we pay a price in reliability for this contextual memory: "Because human memory is so thoroughly driven by cues, rather than location in the brain, we can easily get confused."[4]

Understanding how memory works can have practical payoffs for people who need to learn large amounts of material for school or work. There seems to be a consensus among experts that multiple short study sessions are more effective than are marathon "cramming" sessions just before an exam. Two study periods on consecutive days, for example, is likely to result in *twice* the amount of learning as a single study period of the same total time duration.[5]

If you want to enhance your ability to remember new information, you might also give some thought to what you do immediately *after* you learn it. I read an intriguing study that showed how wakeful resting, simply closing one's eyes and relaxing for a brief period, after a study session can help one better retain the information than if she or he immediately moved on to other activities. This is invaluable information for those of us living in increasingly busy, noisy, and information-soaked societies. Anyone studying for a final exam or preparing for a presentation at work might be wise to resist turning on the television or texting friends the instant studying is completed. Instead, lie back, close your eyes, and do nothing for at least ten minutes. Give your brain a chance to finish its neural processes and lay down those memories in a way that will make them available to you next week, next month, and next year. Don't waste your effort. "Indeed," explains psychological scientist Michaela Dewar, "our work demonstrates that activities that we are engaged in for the first few minutes after learning new information really affect how well we remember this information after a week."[6] Finally, if one wants or needs to remember something important, *write it down as soon as possible and in as much detail possible.* Offload vital information, appointments, or responsibilities to a sheet of paper or a digital document. This is an incredibly simple and effective way to safeguard ourselves against problems stemming from inaccurate memories, yet few of us do it enough.

We also need to be fully aware of what it means for us that our memory is constructive in the sense that recollections of past experiences are assembled from bits of information. It is crucial for us to understand that our brains *tell stories* about past events. These stories may be based on sights, feelings, smells, things heard, and so on. But they must never be thought of as completely faithful transcripts or replays. Neither a memory's clarity nor our confidence about it means much. Don't let that popular video-camera concept creep into your mind. It's wrong, and it leads to unwarranted confidence in memories. There is no recording to play, only a story to tell. Human memories about past experiences come to us in a way that is similar to how archaeologists use bits of information—artifacts— to construct stories about past people and events.[7] "To build a truly reliable memory, fit for the requirements of human deliberate reasoning, evolution would have had to start over," explains Marcus. "And, despite its power and elegance, that's the one thing evolution just can't do."[8]

To make matters even more challenging, personality and self-perception can influence our memories. In addition to supplying us with customized memories intended to assist with present challenges and future planning, our memory system is also set up to help us feel okay about ourselves whenever possible. This can lead our brains to distort particular memories or make them unavailable to us. For example, if you happen to be an extrovert who prides himself on being popular, attractive, and funny, you may not recall that one party in high school when nobody laughed at your jokes and everyone drifted away from you because someone cooler and better-looking was dominating center stage.

In defense of the brain, it is important to point out that the stories our brain tells us about our past may be incomplete, imperfect, or scarcely reality based, but they were never meant to be otherwise. Our memory system works well enough for us during most of our waking hours, and that's the point. The primary goal has always been surviving and functioning well in the real world. The brain has never been concerned with providing total accuracy, because it's unnecessary most of the time. To survive, our ancestors needed to remember the important things, like where predator cats like to ambush Australopithecines, not trivial details such as the cats' eye color. Your brain is not a courtroom stenographer. It is a make-do, can-do organ with limited capacity and a penchant for down-and-dirty efficiency. It doesn't supply you with every bit of data your senses took in during your high-school graduation ceremony because (a) you don't need it and (b) even the magnificent human brain has storage limitations. If we remembered everything about everything, we would drown in meaningless data. Does it really matter what style of shoes the principal was wearing when she handed you your high-school diploma on graduation day? It's not likely to ever be important information so, even if you did notice her shoes that day, the brain lets it go. I have noticed that many brain researchers today don't describe the brain as a warehouse for shelving memories. They emphasize the more creative side of remembering. For instance, "The purpose of human memory is ultimately not to store information but to organize this information in a manner that will be useful in understanding and predicting the events of the world around us," writes UCLA brain researcher Dean Buonomano in his book *Brain Bugs: How the Brain's Flaws Shape Our Lives.*[9]

An aspect of memory that is especially disturbing—and crucial for good thinkers to be aware of—is just how easy it can be to contaminate, alter, or fabricate memories. For example, if a police officer asks a witness to describe the *coat* worn by a murder suspect because it's a cold day and the officer assumes the suspect had been wearing one, the witness might share his or her memory of the coat in great detail, describing the length and color—*even if the murderer wasn't wearing one*. The witness wouldn't necessarily be lying. She or he would "see" the coat in her or his memory because the memory had been altered by a suggestion. A law-enforcement official who understood the risk of influencing recall would make sure to simply ask the witness to describe the suspect and leave it at that. I spoke to a retired police detective about this, and he told me it was a significant concern. He said he was well aware of how easy it was to alter a witness's memories by being careless with the wording of a question. A big worry he had was doing it unwittingly. The best way to get to the truth, he explained, was to hope for multiple witnesses so that a consensus could be found in their stories that likely pointed to something close to what actually happened. For your purposes in daily life, just be aware that this kind of thing is common. If you ask your friend about the time you went to see a particular movie—and include a casual mention of driving to the theater in her car—there is a chance she will remember going in her car, even if the trip had actually been made in your car. I have done this to friends concerning trivial details just so that I could observe this weakness in our memory processes. Never underestimate how easily prompting with words can change memories.

One interesting experiment conducted in the 1970s took this even further and revealed the power of the word "the." Volunteers viewed a short film of a traffic accident and then answered questioned about it. Half were asked if they saw "a" broken taillight and half were asked if they saw "the" broken taillight. Those asked about *the* broken taillight were more likely to remember seeing one. Those asked about *a* broken taillight were less likely to remember seeing one. The word "the" implies there was a broken taillight and the only question is whether or not the witness saw it.[10] Yes, it really can be just that easy to alter a memory.

Another study that also involved video of a traffic accident showed that replacing "contact," "hit" or "bumped" with the word "smashed" when asking witnesses to estimate the speed of the cars caused witnesses to give higher speed estimates.[11] "Smashed" implies a more violent collision, which therefore suggests higher speeds, so that's what many people will suddenly remember having seen.

Work by leading memory researcher and University of California–Irvine Distinguished Professor Elizabeth Loftus has shown that not only can our memories be influenced and tweaked with ease, but it is also possible to have entire memories of events that never happened "implanted" in our brains. No, this does not require a science-fiction scenario involving electronic implants and open-skull surgery. It can be achieved with nothing more than talk or an altered photograph. A memory that feels authentic to us can come to life in the brain, created and fueled by little more than ideas, images, and suggestions. Consider what this says about human memory that you need to be aware of: One person can literally talk another person into remembering something that never happened. The most frightening thing about all of this is that our memories can feel absolutely real and reliable, even if they have been distorted, edited, or embellished by our own brain or are thoroughly false and have been implanted with the help of some external source. Loftus writes that "manufactured memories are indistinguishable from factual memories."[12] A Washington University study that used word lists to influence participants' recall found that people can be better at remembering things that didn't happen than they are remembering things that did happen![13]

The unexpected nature of human memory and its shortcomings in accuracy and reliability constitute vital knowledge. Good thinking certainly is compromised without this awareness; I would argue that it is impossible. "Seeing" the past clearly in one's head simply is not good enough when it comes to important matters based off of past experiences. "Our memories are constructive; [and] they're *reconstructive*," Loftus explained in a 2013 TED Talk. "Memory works a little bit like a *Wikipedia* page. You can go in there and change it—but so can other people. . . . We can't reliably distinguish true memories from false memories. We need independent corroboration."[14]

INACCURATE MEMORIES VS. ACCURATE MEMORIES

A Washington University experiment found that 82 percent of 96 participants remembered reading words on a list of items that shared a common theme *even though those words were not on the list they read*. Fewer students, 75 percent, remembered only words that actually were on the list. This suggests that we are capable of remembering things that our shadow brain assumes *should have* been present or happened—but weren't there or didn't happen—at a higher rate than we remember real things and events.[15]

It is clear that we all will be presented with inaccurate memories many times throughout life. They will come at us often, both from others and from our own brains. Ignoring this or believing otherwise is a recipe for error and delusion. Human memory is messy and unreliable and should not be trusted excessively or exclusively when it comes to important matters.

Overconfidence in the memory of eyewitnesses to alleged crimes has been a significant problem in police departments and courtrooms for a long time. No firm numbers exist, but a 1989 survey of prosecutors estimated that a minimum of 80,000 witnesses identify people as part of law-enforcement investigations each year in the United States.[16] How many of these roughly 80,000 witnesses honestly shared their memories with police, prosecutors, and/or juries but were wrong about the reality of what happened because their memory, like everyone's memory, was fallible? How many jurors have wrongly thought that a witness's confidence is a reliable indication of her or his accuracy? How many innocent people have been wrongly convicted not because they were the victims of lies, corruption, or a mediocre defense attorney but because the legal system did not account for the realities of human memory? According to the Innocence Project, which works to free wrongly convicted prisoners, DNA tests have led to 325 postconviction exonerations in the United States. Twenty of these were death-penalty convictions.[17] What was the most common element in these convictions that were found to be unjust? *Eyewitness identification*. Of the 329 cases, 236 were because of eyewitness misidentification.[18]

The National Academy of the Sciences convened an expert panel to review what is currently known about vision and memory as it relates to problems with eyewitness testimonies. Its incisive 2015 report, *Identifying the Culprit: Assessing Eyewitness Identification*, presents a disturbing picture of the US legal system's past and present use of witnesses. It also pulls no punches in explaining how unreliable witnesses can be at accurately reporting what they see:

> Human vision does not capture a perfect, error-free "trace" of a witnessed event. What an individual actually perceives can be heavily influenced by bias and expectations derived from cultural factors, behavioral goals, emotions, and prior experiences with the world. . . . As time passes, memories become less stable. In addition, suggestion and the exposure to new information may influence and distort what the individual believes she or he has seen. Several factors are known to affect the fidelity of visual perception and the integrity of memory. . . . The recognition of one person by another—a seemingly commonplace and unremarkable everyday occurrence—involves complex processes that are limited by noise [random interference] and subject to many extraneous influences.[19]

Although the challenges facing the legal systems of America and the world are vast and complex, the report's authors conclude that significant improvements can be made with broad implementation of changes, including these:

- Double-blind lineups and photo arrays. This means that the investigator interacting with a witness would not know if a particular suspect or arrested person is in the lineup and therefore couldn't knowingly or unknowingly influence the witness to select that person.
- All law-enforcement officers should be trained in how to interview and interact with eyewitnesses. This would help to reduce the potential for unwittingly altering or contaminating witness memories.
- There should be standardized instructions given to all witnesses.
- More scientific *specialized* research is needed that relates directly to the use of eyewitnesses in law enforcement. Pure research efforts do not always transfer well to the criminal-law environment.[20]

The world's legal systems need to catch up to what science has already revealed about human vision and memory, as well as pay close attention to new discoveries. Fortunately, this is happening in some places. In 2011, the New Jersey State Supreme Court introduced sweeping changes to how eyewitness testimonies are handled in courtrooms, including a requirement for judges to explain to juries that there are many problems associated with vision and memory if questions were raised about reliability during the proceedings. Chief Justice Stuart J. Rabner wrote in his opinion: "Study after study revealed a troubling lack of reliability in eyewitness identifications. From social science research to the review of actual police lineups, from laboratory experiments to DNA exonerations, the record proves that the possibility of mistaken identification is real. . . . Indeed, it is now widely known that eyewitness misidentification is the leading cause of wrongful convictions across the country."[21]

One year later, the Court went further and issued instructions to all New Jersey judges that they *must* inform jurors about some of the potential problems with eyewitness testimonies before deliberations begin. The instructions include these statements: "Human memory is not foolproof. Research has revealed that human memory is not like a video recording that a witness need only replay to remember what happened. Memory is far more complex."[22]

This particular intersection of law and brain science is of great importance and is sure to continue shaking the foundations of legal systems everywhere for many years to come. It needs to because it is clear that police departments and courts have relied too heavily on the fragile memories of eyewitnesses when searching for suspects and pursuing convictions. Given the high stakes of crime and punishment, the need for change is obvious. Furthermore, the lesson here for those who would dedicate themselves to good thinking is profound. If honest, intelligent human beings err routinely when seeing and recalling dramatic crimes—even when they themselves are the victims—how can any of us trust our own less important memories? Perhaps our memories might best be thought of as a starting point, a place from which to begin our search for answers and wise decision making.

HOW WE FORGET THE MOMENTS WE'LL NEVER FORGET

Where were you when the 9/11 attacks occurred? Many people have vivid memories of September 11, 2001, because of the event's dramatic nature and the extensive television-news coverage. I certainly remember the moment when I first heard about what was happening in Manhattan. I was having one of those rare days in life when everything flows like a fairy tale. I was in Florida, at Disney's Animal Kingdom theme park with my wife and young children. I was about as far away as one can get from the hate and violence that too often sours our world. When an employee told me about planes hitting the World Trade Center towers, however, all those problems came crashing into my imagined sanctuary. My memories of that day still contain a lot of detail and emotional flavor, more than a decade later.

It was an ideal morning, clear and sunny but not too hot. The park was busy but not uncomfortably crowded. I still remember the face of the female employee who told me about the buildings being hit. I watched an older woman cry hysterically as she made her way through the crowd. I assumed she was reacting to the news. Maybe she feared she had lost someone close to her. Concerned parents led confused and protesting children toward the park exit. Two plain-clothes security people, a man and a woman, lingered close by and watched over guests at the shuttle pickup area. Their radios were poorly hidden in Mickey Mouse shopping bags. The male security agent rubbed his chin while scanning for threats. My son looked to me for answers that morning: "Why do we have to leave so early, Daddy?" I can't remember answering his question. Most likely because I didn't. Three-year-old children don't need to have everything explained to them. I remember looking out the window of the bus as we rolled away on our way to the hotel. I felt sad, frustrated, irritated. Some of the world's hate and madness had penetrated, spoiled, and shut down the place Disney's marketers call the "happiest place on Earth." I remember thinking how unfair it was for my children to be denied that day of happiness. I had not seen any television coverage yet, so I couldn't fully grasp the enormity of the tragedy. Looking back on it now, however, I am ashamed of myself. It was selfish and shallow to grieve over a spoiled vacation day, given all of the death and suffering that occurred that day. But that was how I felt at the time.

Here's the thing about my crystal-clear, detail-laden recollection of the morning of September 11, 2001: I don't trust it. I know enough about memories, even special and highly emotional ones, to be skeptical of any and all details. Sure, it feels completely accurate to me. I can "see" so much of that morning in my head. But I can't be absolutely sure about most of it. My memory system might easily have added elements that did not happen to me or left out important things that did. The only reason I am completely confident about even a fact as basic as where I was that day is because my wife was with me and her memories corroborate it. I also have photos. Contrary to common belief, multiple studies have shown that even these emotionally charged "flashbulb memories," as psychologists call them, are subject to significant decay and distortion.

Several studies utilized the 9/11 attacks to learn how intense memories hold up over time. Researchers asked college students to write down brief answers to basic questions the day after the attacks, including where they were when they first heard what happened, whom they were with, and what they were doing immediately prior to learning about the attacks. By revisiting the same questions with the test subjects weeks and then months later, and then comparing statements, researchers found that the accuracy of a flashbulb memory deteriorates over time *just like ordinary, everyday memories*. The only difference between a typical flashbulb memory and an everyday memory is that a person's *confidence* in the accuracy of a flashbulb memory tends to remain very high. So, I'm sorry to say, even the sharpest recollections of the most emotionally intense and important events of our lives can be and likely are riddled with errors.[23] But what if I had been in downtown Manhattan that morning instead of at a Disney theme park?

Proximity to an emotionally charged event and personal danger are key to a flashbulb memory's staying power, and possibly its accuracy over time, because of the involvement of the amygdalae (the brain's fear center). My amygdalae were probably not highly active on 9/11. Other than the vague possibility that Disney parks were secondary targets, I was never in any physical danger. My experience was in the form of public trauma. I felt my mix of fear and confusion inside a community of vacationers, and my greatest horror came from only imagining what others had felt in those buildings and on those planes. But what if I had been near Ground Zero? No doubt my amygdalae would have been ablaze with activity, and my memories today would likely be more intense and perhaps more accurate as well.

A fascinating 2004 study used fMRI (functional magnetic resonance imaging) to scan the brains of people as they recalled the 9/11 attacks three years later. All of them showed activity in the hippocampus while retrieving the memories, which is how most everyday memories are sorted and retrieved. But those who had been downtown, near the World Trade Center, showed significant activity in their amygdalae as well.[24]

Jon Simons, a cognitive neuroscientist at the University of Cambridge, offers this practical advice for decision making with brains that cannot always be trusted to remember key details:

> For big decisions that may have serious implications in our lives, it's clearly important to ensure we take the time to evaluate properly the advantages and disadvantages of each course of action available to us. Unfortunately, for many real life decisions it can be difficult to quantify the pros and cons of each alternative, so we tend to rely on heuristics, rules of thumb that are often good enough in most situations. These heuristics often depend on remembering how we dealt

successfully with similar problems in the past, so accurate memory for those previous situations is important. My advice for using heuristics to make any weighty decision with serious implications is to seek out documentary evidence of what worked last time, rather than relying on your possibly fallible memory![25]

Good thinking demands plenty of caution and humility regarding our reliance on memories. Confidence and detail alone are not enough to validate a memory. Who you are does not matter on this point. Even master skeptics and committed critical thinkers face great dangers when the stakes are high. Caution, doubt, and careful review are always wise if a memory is to be a key factor in making an important decision or trusting an extraordinary claim. Good thinkers may forget much, but they do remember the limitations and problems that come with human memory. We all have what amounts to a production team inside our brains. That team presents us with versions of reality, *docudramas* as the episodes of our lives. Thanks to science, we know that memory is something like our own personal *reality TV show*. It is supposed to be real, and much of it is, but much of it is scripted fantasy, too. We have been warned.

Chapter 7

THE SHADOW BRAIN

"He felt all at once like an ineffectual moth, fluttering at the windowpane of reality, dimly seeing it from the outside."

—Philip K. Dick, *Ubik*

Most of us have heard of something called "the subconscious mind," that mysterious *other* part of ourselves that somehow interacts with our normal mind. What few people grasp, however, is just how prominent and powerful an influence it is to our lives. This subconscious presence, the shadow brain, is not a minor player who follows behind you only to occasionally whisper advice in your ear. No, it is you who follows the shadow most of the time. Believe it or not, "you" are the minor player in your life.

To help understand how important the shadow brain is to you, imagine looking down from high above and spotting a lone swimmer treading water at the center of an enormous lake. The swimmer is you, and the lake is your shadow brain. The swimmer is tiny and alone, with deep waters on all sides. The swimmer is also blindfolded. Currents and waves constantly pull and push one way or another. The blinded swimmer cannot consistently detect or make sense of these forces because there are no points of reference. Sometimes the currents turn the swimmer a bit left or right. Sometimes they completely reverse the swimmer's course. Often the swimmer will stroke away at full effort but go nowhere. Sometimes the oblivious swimmer doesn't swim at all but makes fast progress across the lake nonetheless due to strong currents. In addition to this, unknown creatures swim around and beneath the swimmer. Sometimes they brush against the swimmer's legs with such a light touch that the swimmer doesn't notice. Some are small, some big. Many times they impact with such force that the swimmer knows something substantial is there but still has no idea what it is or what its intentions are.

The swimmer has no idea how wide or how deep the lake is. The shore could be ten meters away or ten thousand kilometers away. But the swimmer imagines it is nothing more than a small swimming pool because it is a comforting thought. The lake is a thousand times larger than that, however. If only the swimmer knew more about the lake, at least a hint of its size and maybe something about the currents, waves, and creatures as well. But the blindfold prevents the swimmer from understanding anything. In an attempt to make sense of it all and feel less anxiety, the swimmer constantly thinks up reasons, explanations, possibilities, and excuses for why the waters move, why strange things move around below the surface, and why no shore is ever reached. The swimmer keeps swimming, year after year, never realizing the bizarre truth: The lake is the greater part of the swimmer's life. The lake is the swimmer, too.

Good thinking can't take us out of the vast lake that is our shadow brain. But that's okay. We wouldn't want to leave it even if that were possible. We need our shadow brains to be our automatic-reaction force, keep our involuntary systems functioning, and to find countless mental shortcuts through daily life for us. But good thinking does one invaluable thing for us. It takes off the blindfold. Good thinking lets us understand that we are that swimmer and that the lake is us too, influencing and controlling us. No longer blind and ignorant, we can now pause to reason when it is appropriate and possible, such as the moment before we make important decisions. We can pay attention and try to determine if we swam a northerly course for sensible reasons or if it was one of the creatures below that nudged us northward for irrational or unknown reasons. We will always be in the center of the lake; the lake is us as much or more than anything else in our lives. We will always find ourselves treading water and swimming with or against the currents of our subconscious. Simply recognizing this reality, however, gives us greater control and provides more opportunities to make good decisions.

ARE YOU IN CONTROL OF YOU?

It is difficult to overstate the influence and impact of the shadow brain on one's life. You don't even live your life in real time. It typically takes a fraction of a second for your shadow brain to receive, process, and act on input from your senses, significantly faster than your conscious mind. So when a friend yells your name from across the street, a cat rubs against your leg, or your phone rings, "you" are the last to know. Weirder still, your *other* brain is first to act even when "you" decide to do something! Researchers have observed related brain activity occurring *before* a person makes the conscious decision to act.[1] This means that at the precise moment you decide to stand up, for example, your shadow brain has already begun the cerebral processes related to standing up. It knew what you were going to decide and started working on it before "you" actually made the decision. What does this imply about free will? If your shadow brain has already decided to stand up before "you" did, then who is really doing the standing up? Who is in charge? Moreover, this "other you" that is so involved with your life is constantly feeding you input about how to react to the things, events, and people you encounter. Should you buy this or that? Is this person I just met okay or a bit too creepy? Trust me, while you might want to dither and ponder ethics or weigh a long list of pros and cons, the other you has already made the call. More often than not, "you" are relegated to the role of *explainer in chief*. You have to defend and make sense of decisions and actions your shadow brain is responsible for. "To make our way in the world," warns Yale psychologist John A. Bargh, "we must learn to come to terms with our unconscious self."[2] Otherwise we can do little better than flounder in the middle of a deep lake—blindfolded.

While in pursuit of good thinking, we have to acknowledge that the shadow brain is always there, looking over our shoulder, paying attention when we are not, working through problems we gave up on, and making countless split-second decisions that we then take credit or blame for. It may be common for us to second-guess the words and deeds of others. But modern science has made it clear that we should be in the habit of second-guessing ourselves as well. Neuroscientist David Eagleman, director of the Laboratory of Perception and Action at Baylor College of Medicine, doesn't hold back in describing the unbalanced roles of our two minds: "The conscious mind is not at the center of the action in the brain; instead, it is far out on a distant edge, hearing but whispers of the activity."[3]

It is beginning to seem as if we are little more than bystanders in our own lives. Psychologist Bargh suggests that we let go of the notion that the subconscious is some mist-shrouded, diluted version of ourselves prone to momentary visits.

> Freud spent countless thousands of words in providing explanations as to why our unfulfilled wishes express themselves in the imagery and stories that populate our nightly dreams. The latest research provides a more pragmatic perspective on how thought and emotion just below the surface of our awareness shape the way we relate to a boss, parent, spouse or child. That means we can set aside antiquated notions of Oedipus complexes and accept the reality that the unconscious asserts its presence in every moment of our lives, when we are fully awake as well as when we are absorbed into the depths of a dream.[4]

If you are beginning to feel uncomfortable right about now, don't worry, this is a common symptom of new and meaningful education. It will pass in time.

When we make important decisions—choosing which candidate to vote for, investing our money, hiring an employee, trusting someone with a vital secret—it may feel to us like we are in total control and coming to rational and fair conclusions based solely on logic and facts. In reality, however, the shadow brain is always there, nudging, cajoling, shoving, or demanding action in response to its hunch, one that is often based on nothing more than a quick visual assessment of someone's facial symmetry, height, social rank, or skin color. Unless the job is modeling, most people are likely to say that hiring applicants based on how well the left and right sides of their faces match up, down to the millimeter, is not only ludicrous but also unfair, unproductive, and perhaps even evil. But it happens every day because facial symmetry is the most important factor in physical beauty and the shadow brain is a sucker for good looks. And while well-aligned faces and bodies may translate to our subconscious as indicators of health and genetic fitness that are important for potential mate selection,[5] facial and body symmetry certainly should not come into play during the process of making most of our decisions and transactions. Yet you have been nice or indifferent many times to people, bought things you didn't need, and agreed or disagreed with people based on their cheekbone and eyebrow alignment. You have done it, and I have done it, because the shadow brain told us to.

To be clear, the subconscious mind is not some sort of evil demon we desperately need to cast out of our bodies. The shadow brain should not be thought of as the enemy. It is not on a mission to trip us up and make fools of us at every opportunity. To the contrary, its purpose is to help, to keep us safe and make us successful. And it serves us well most of the time. Which only makes sense because, remember, the shadow brain is "us" as much as our conscious mind is. The lake is there to serve and protect, not drown, the swimmer. If you are walking on a street and hear the faint sound of a dog growling up ahead, your conscious mind may reflect on how much you love dogs and how sweet they can be. But your subconscious mind might feel differently and be quick to remind you that growling dogs sometimes bite, and, before you know it, you are crossing to the other side of the street. Your shadow brain may just have saved you from a series of painful rabies shots. Sometimes we get a bad feeling about someone for good reason. "You" may see nothing wrong with the new neighbor and invite him over for dinner to welcome him to the neighborhood. Meanwhile, your shadow brain might pick up on subtle cues that you missed and warn you that he is definitely a serial killer. So even if we could tune out its input, why would we want to? It certainly makes more good decisions than bad for us. If it didn't, we would have gone extinct long ago.

Question everything, your own thoughts, perceptions, and decisions especially. But do not fear or fight the shadow brain as a matter of routine. We will never be Vulcans or robots. Well, on second thought, the latter might happen, but that's another discussion for another book. What I am calling for is more oversight between "you" and your shadow brain. Acknowledge its constant presence and dominant role in your life. Remember that our shadow brains deliver a steady barrage of instant hunches that our conscious minds then work to justify and defend. Make peace with the situation, and then do your best to review the shadow brain's demands and recommendations whenever possible in the light of reason. Simply knowing this and keeping it in mind throughout daily life can prevent many mistakes over a lifetime. The key is to challenge your shadow brain and make it defend its positions.

Am I about to join this organization because I agree with what its members do, what they stand for, and because it will benefit my life? Or could it be because the person recruiting me is beautiful, authoritative, or compelling to me in some irrelevant way? Why am I about to buy this herbal supplement? Is it because I want to believe in it, because a friend who is not a doctor told me it works, or do I just like the attractive packaging and wording on the box? Why have I decided to cast my vote for this candidate? What are my objective reasons? Is it possible that I'm not thinking and I'm just being loyal to a political party above my own interests and that of my society? Has confirmation bias and motivated reasoning taken me down an irrational path? Have I been swayed by misleading campaign ads? Has my assumption that this candidate seems trustworthy in TV interviews distracted me from the fact that I disagree with her position on several issues?

When we do our best to place input from the shadow brain—which tends to be fast and sloppy with facts, evidence, and logic—alongside input from our (hopefully) rational conscious self, we position ourselves to make better decisions most of the time. "You" may be a minority, unable to take control of the murky universe inside your head, but that's okay. At least you can have more of a say in what goes on. Just make sure to understand that your brain and body are put together in a way that necessitates the shadow brain doing most of the work. Neuroscientist Eagleman describes the state of our brains in his book, *Incognito: The Secret Lives of the Brain*:

> Brains are in the business of gathering information and steering behavior appropriately. It doesn't matter whether consciousness is involved in the decision making. And most of the time, it's not. Whether we're talking about dilated eyes, jealousy, attraction, the love of fatty foods, or the great idea you had last week, consciousness is the smallest player in the operations of the brain. Our brains run mostly on autopilot, and the conscious mind has little access to the giant and mysterious factory that runs below it. . . . Our neural circuits were carved by natural selection to solve problems that our ancestors faced during our species' evolutionary history. Your brain has been molded by evolutionary pressures just as your spleen and eyes have been. And so has your consciousness. Consciousness developed because it was advantageous, but advantageous only in limited amounts.[6]

Do not imagine that it is easy to challenge questionable or outright bad suggestions and commands fed to us by the subconscious mind. First of all, it's difficult, if not impossible, just to realize that it's happening. By definition, these inputs originate apart from our consciousness, meaning we are not fully aware of them, if at all. It's a bit like trying to joust with ghosts. We are at a disadvantage. Through attentiveness, however, we all can at least keep an eye on our decision-making process enough to catch many blunders in the making before it's too late. This is about averages over years, over a lifetime. None of us will catch every silly or dangerous thought that bubbles up from the subconscious. But as good thinkers we certainly can detect and override enough of them to make ourselves safer and more productive, while wasting less time and money.

THE SHADOW BRAIN GOES SHOPPING

When contemplating a purchase, whether it is a pair of shoes or a new car, be sure to routinely pause and practice good thinking during the process. You might ask yourself why it is that you are drawn to one product over all the others. Are you being manipulated somehow? If you are standing in a store, viewing an online merchant's website, or looking at an ad, then the answer to that question is almost always *yes*. Maybe a natural cognitive bias called the *framing effect* has focused your attention on the more profitable item the store wants you to buy, a transaction that would be good for them but not for you. This is easily done by positioning the target product next to a similar but higher-priced item. Even restaurants do this with their menus. The price comparison will make you feel better about the price of the less expensive item, which makes you more likely to buy. Be alert to these situations and assess the product in isolation. It's not a foolproof method, but imagine if the item you are interested in wasn't next to a pricier option. Does it still feel like a good deal for you? If not, walk away.

Remember that emotions override reason with ease. Try to predict how you are likely to feel about the item six months later, a year later. Are you allowing your shadow brain to talk you into trading cash for a momentary distraction that you will forget or regret tomorrow? Or is this something you genuinely need or want in a meaningful way? Adopting the habit of a pause and brief chat with yourself is likely to save you many thousands of dollars over a lifetime. Before you toss that shiny trinket into your shopping basket or say yes to that grinning salesperson with the highly symmetrical eyes, *think*. Who really is making this decision? Is it "you" or your shadow brain? Admit in this crucial moment that you are an irrational being who is easily pushed around by the impulsive shadow brain. Acknowledge that your subconscious mind may not be considering deeper implications of this decision, such as your long-term financial future. It likely is driven by an immediate want, belief, fear, or hope. You can be sure it hasn't paused to carefully assess evidence or weigh the pros and cons. That's your job. It is also important to understand that the pre-purchase thoughtfulness I am encouraging is not the same as knowing information about a product or service. That can matter too, of course, but this is about you and what is going on inside your head. Too many shoppers and investors make great efforts to research products, services, and companies. But all that work means little without good thinking. We must look within and analyze ourselves, too.

Sometimes a minimal amount of conscious observation is all that is required to improve one's decision-making skills and thus save substantial amounts of money. *Look* at the product you are contemplating buying. *Read* the basic information that the manufacturer has presented to you. Mere seconds are often enough. In many cases, everything you need to know to make a rational decision is right there in front of you. But you are likely to miss it if you have allowed the shadow brain's rapid reactions based on beliefs and cravings to effectively blur your vision or put your reasoning skills to sleep. For example, during one of my medical-quackery safaris, I encountered copper bracelets for sale at a drugstore in California. Printed on the front of the packaging was the enticing statement that copper bracelets "have been worn since Roman times in the belief that they are instrumental in affording relief [for] ARTHRITIC & RHEUMATIC conditions." (The two medical conditions are presented in all caps in the original text.) Sounds pretty good, right? Someone who suffers from arthritis pain and feels favorably about alternative medicine is probably going to buy a couple of these. Maybe pick one up for the dog, too. But wait, on the back of the package, this is what I found: "This copper bracelet is not being sold as a medical device

& not intended to diagnose, treat, cure or prevent any disease." A reveal like this is more common than you might think. Perhaps it is required in some cases or is done voluntarily out of a concern for potential legal troubles, but many companies do state in plain language on their own packaging or in their ads that their product is junk and not worth the money they are asking for it. In many cases, consumers only have to take a few seconds to look. Unfortunately, hopes and beliefs from the shadow brain can make us miss, dismiss, or quickly forget telling statements that should make us be more rational and leave us with more money in our wallets.

Remember Erik von Däniken's book, *Chariots of the Gods*, mentioned in chapter 1? Although it is usually shelved in the nonfiction sections of bookstores and libraries, categorized with nonfiction astronomy and archaeology books by online booksellers, and taken to be a factual work of scientific research by millions of people, it has this unambiguous statement printed on the copyright page: "This is a work of fiction."[7]

LOOKING FOR PATTERNS IN ALL THE WRONG PLACES

Former president of Iraq Saddam Hussein was executed by hanging on December 30, 2006. Although many applauded the event as well-deserved justice for a brutal dictator, others felt different. There were heartbroken Iraqis who had remained loyal to Saddam through the United States–led invasion and were now afraid of what their country would become without him. During this time of tumultuous transition and high emotions, some Iraqis looked up and saw their lost leader. The face of Saddam Hussein had taken over the Moon. Iraqi citizen Marua Diya told an NPR reporter what he saw: "Saddam's image was very clear—his face, his glasses. His face was there on the Moon."[8] Were these people who claimed to see Saddam in space insane? Nope, just human. That's all. The shadow brain not only whispers numerous absurdities to us over a lifetime but also shows us more than a few.

In the 1990s, a Florida woman made a grilled-cheese sandwich in her kitchen, took one bite, and then saw a face staring back at her. She recognized the person immediately. It was Mary, the virgin mother of Jesus. With great care, the woman placed the sandwich in a clear, plastic box and surrounded it with cotton balls for protection. She kept it on a nightstand by her bed for the next ten years before deciding that it was her duty to share the grilled-cheese sandwich with the world. She put it up for auction on eBay, where it was purchased for $28,000.[9]

She's not alone. In recent years Jesus's face has been spotted on a tortilla,[10] a potato chip,[11] the wings of a moth,[12] and even a fish stick.[13] Wouldn't you love to ask these people how they knew it was Jesus or the Virgin Mary they saw? After all, no one really knows what any of these religious figures are supposed to have looked like. All one would have to go on is the work of painters and sculptors, none of whom ever saw either one of them.

This is not limited to central figures of Christianity, but because physical depictions of Islam's prophet are forbidden within the religion and too often lead to violence, one never hears of his face turning up in food. However, the BBC reported in 2004 that Mohammed's *name* was discovered written in Arabic on the side of a young goat.[14] Over the years, I have read many accounts of people who "saw" Allah's name written in bread. Allah's name can be heard in unexpected places as well, say believers. I once saw a short video of a lion in a zoo "saying" Allah's name.[15]

None of this is new or rare, of course. Believers worldwide have been claiming to see, hear, and feel the objects of their belief in a variety of places for thousands of years. Different religions, same claims. We also can safely assume that humans have been doing this far back into prehistory because science has shown us how the shadow brain naturally misleads us to see and hear things that are not there. This is not an experience that is limited to religious believers, of course. In 2015, a hotel owner in Scotland chopped down a tree on his property and immediately recognized the face of ET, star of the 1982 film *E.T. the Extra-Terrestrial*, staring back at him from the stump.[16] Thankfully, everyone involved seemed to take this one to be a cute coincidence rather than a supernatural event or a literally planted message from an alien species. The fact that this kind of thing happens routinely and in virtually any context indicates that the explanation for it is likely to be found somewhere inside the human brain rather than outside it, in so-called supernatural or paranormal realms. The shadow brain is very good at utilizing the already-strange vision system to spot things we want to see, regardless of whether or not they are really there. We are always on the lookout for meaningful patterns in meaningless noise; we are driven to make something out of nothing. The good thinker needs to understand and remember this because it has implications for everything from traffic and hunting accidents to UFO and ghost encounters. It also influences the way in which we tend to think about the universe and ourselves. Paleontologist Stephen Jay Gould recognized the central significance of our pattern-seeking compulsion: "No other habit of thought lies so deeply within the soul of a small creature trying to make sense of a complex world not constructed for it."[17] The shadow brain is relentless in its quest to feed you important information as fast as possible. In its haste, however, mistakes will be made. The process often involves latching on to information already in the brain to help make sense of new input. This is one more reason to keep your head as clear of nonsense as much as possible. Allowing yourself to embrace an irrational belief out of desire or laziness may come back to haunt you in the form of sensory mischief that plays to the baggage in your brain.

Psychology professor Hank Davis appreciates the good that comes from our pattern-seeking compulsions but offers an important warning too:

> The human mind is an illusion generator. Since you have a human mind, that statement probably either feels wrong or like an insult. It is neither. Patterns are everything to us. We hunger for them. We revel in them. They are the basis for art, literature, music, and much more in our lives. But a perceptual system that is so geared to wrestling patterns out of complex arrays of stimuli is bound to produce some false positives. From time to time we are going to see or hear what is not there, and those cases will seem no less compelling to us.[18]

So why did we end up with this odd ability to find dragons in clouds, alien spaceships in the sky, and even gods in our food? Why does every normal, healthy human brain come preloaded from the factory with this high-speed, automatic, overzealous pattern-recognition capability? Why should we be so good at finding and picking out a meaningful image or sound that is lost within a clutter of visual or auditory information? It's easy to recognize why when we think about the environment that shaped us. For most of our existence we were constant combatants in the brutal game of competition between predator and prey. We were on the menu of some life-forms and some life-forms were on our menu. Simply being seen, or not, is a critical factor in the struggle for survival. Visual detection comes before all the running and fighting. This is why the leopard has spots, the chameleon changes color, and the moth has false eyes on its wings.

If we were no good at noticing a predator waiting in the bushes during our daily strolls, then we would have stumbled into extinction long ago. We never would have made it to *Homo erectus*, much less *Homo sapiens sapiens*. Our ancestors' survival depended not only on spotting the big cats intent on eating them but also on being able to identify a nest full of protein-rich bird's eggs, insects, tasty fruits and berries hidden among the leaves, well-camouflaged rodents in the weeds, and so on. If they couldn't do it, we wouldn't be here. Therefore, the ability to see something that was difficult to detect was a crucial ability, a prerequisite to modern humanity. I believe that the difficulty of this constant challenge our ancestors faced led us not only to be very good at pattern recognition but also to be downright obsessed with it. Outside of microbial life, camouflage is the most common defensive and offensive survival strategy in all of nature, and this was a huge evolutionary influence on our brains. The problem today is that so few people are aware that their shadow brains are always searching for patterns. You and I, here in the modern world, are working with a prehistoric brain that is always on. It never stops looking, listening, and thinking in an attempt to identify the meaning within what it cautiously perceives as deceptive meaninglessness all around us. It is on an important mission to identify hidden dangers and opportunities. Sometimes it tries too hard. It is as if our brains are perpetually suspicious, afraid, and hungry. And that's why we are so vulnerable to making the mistake of sensing things that do not exist.

This phenomenon of connecting dots that have no meaningful connections may be referred to as *apophenia* or *pareidolia*. Michael Shermer, science historian and founding publisher of *Skeptic* magazine, calls it "patternicity," which he defines as follows: *The tendency to find meaningful patterns in both meaningful and meaningless data*.[19] My minor quibble with that is that the word *tendency* makes this seem way too passive and infrequent.

After many years of listening to people on six continents tell me about their miracles, visions, magical healings, and paranormal encounters, I would describe pareidolia as nothing less than a full-blown universal *compulsion*. It's as natural as breathing, I suspect. Shermer adds *agenticity* to his patternicity. This is our tendency (compulsion?) to give meaning to these patterns we keep detecting and to imagine intelligent control behind it all. "We believe that these intentional agents control the world," Shermer explains, "sometimes invisibly from the top down, as opposed to bottom-up causal randomness. Together patternicity and agenticity form the cognitive basis of shamanism, paganism, animism, polytheism, monotheism, and all modes of Old and New Age spiritualisms."[20] Given the frequency and consistency of extraordinary claims about seeing, hearing, and feeling things that almost certainly are not there, it seems that we just cannot help ourselves. It is as if our brains are determined to find something, *anything*—no matter what we look at, listen to, or think about—and then inject it with meaning that comforts us, reinforces a current belief, or at least makes some degree of sense to us. Which is why good thinking—specifically skepticism and the scientific process—is vital. It's the best way we have to sort through all the suggestions and conclusions coming at us from our overactive and rapid-fire shadow brains and attempt to distinguish what is real from what is not.

Our obsession with searching for patterns in everything is not limited to sights and sounds, of course. We also do it with facts. If I flip a coin and by chance it comes up heads five times in a row, how much money would you bet on me getting heads again? Some might feel that tails are due, so that's the smart bet. Others might figure that I'm on a streak and heads is the right call. The actual odds? It's fifty-fifty, of course. Each flip of a coin is independent of previous flips. But your shadow brain likely would try to convince you otherwise because it would imagine a meaningful pattern in all those flips. This is why "you" must question, doubt, review, reconsider, and think before betting, buying, joining, or doing anything else with significant consequences. Good thinking is the crucial second pass over your shadow brain's ideas and conclusions. Skip it at your own peril.

Chapter 8

THE ENEMY WITHIN

"It is never too late to give up our prejudices."
–Henry David Thoreau, *Walden*

I once attended a tense Ku Klux Klan rally in Brooksville, Florida—as a journalist, not a member. The hate was thick and loud that day. "We want n—s to go back to Africa! It's best for them and it's best for America! N—s belong in jungles, not in the USA! This country was built by white people, for white people!" These colorful declarations came from an angry man with a bullhorn. He stood on the steps of the county courthouse and a statue of a Confederate soldier stood guard nearby. A wall of armored police separated some one hundred anti-Klan protesters from him and the rest of the KKK members. Many of the black protestors shouted, "Black power," in an attempt to beat him with volume. As this went on for more than an hour, thoughts of past crimes and future struggles for humankind swirled in my mind.

On another day, I visited the headquarters of the African People's Socialist Party in St. Petersburg, Florida. One of the first things I saw upon entering the building was a black-and-white print on the wall that showed a black man hanging from a tree. The rope had stretched his neck to grotesque proportions. A crowd of gleeful white people were gathered below.

I interviewed one of the leaders of this black revolutionary organization. Polite but blunt, he explained to me that their goal was to establish a separate nation for black people somewhere within North America—no whites allowed. Their path did not allow for any reconciliation or blurring of human borders. "We have to consider violence as the way to improving the situation of black people in America," he said. "They [white people] have all the power, money, and resources. We cannot imagine them giving it up peacefully."

On yet another occasion, I spent the better part of what should have been a pleasant and sunny afternoon with a gathering of Neo-Nazis in a public park. That weird mash-up of a picnic, hate rally, and World War II reenactment left me wondering if I had been born on the wrong planet. I heard more repulsive quotes about human beings that day than I care to remember.

The racism I observed, heard, and felt during those three encounters had one thing going for it. At least it was identifiable and undeniable. Loud, in-your-face, and boorish racism is relatively easy for the more fair-minded and enlightened among us to counter, push back against, or quarantine. But what if racism and other destructive prejudices come without banners, flyers, and hate mongers screaming into bullhorns? How do we resist when the hate or fear is stealthy, silent, and everywhere? And what if some of it is even within you, lurking somewhere within your shadow brain? What then?

In my book *Race and Reality: What Everyone Should Know about Our Biological Diversity*, I survey many of the logical and scientific problems with the popular concept of biological race categories and explore how it negatively impacts social harmony, education, sports, and more. Packaged and sold as a reality nature imposed upon us, races are in reality culturally constructed groups that we can only blame ourselves for. We know this because the race rules of inclusion and exclusion vary radically across cultures as well as continually change over time. Biological diversity is real, of course, but it does not fracture us into that collection of a few tribes writ large that most people still believe in. The claim that there are three, five, or several biological races with meaningful borders around them fails in the light of scientific evidence and observable realities. One can make a much better argument for there being many millions of biological races or, even better, none. The American Anthropological Association issued a concise statement on the problems with race belief in 1998 that is well worth reading.[1] In the face of so much violence, hatred, and inequality, we must recognize that it could only be better for the world if more people had a realistic perception of humankind's biological diversity and where they fit into it according to science. But maybe reading a book or a statement by a bunch of anthropologists is just not enough. Maybe some prejudices against groups of people are deeper and more difficult to address.

Chances are you are not a rabid racist who spends your waking hours despising and plotting the downfall of all those misfortunate enough to have been born to a "race" other than yours. Such people do exist, of course, but I'm willing to assume that you are not one of them. Thanks to your modern education and socialization, you probably believe everyone should be treated fairly and that tolerance is better than intolerance in most cases. And although there are still race problems everywhere and some societies continue to struggle more than others, at least we can identify a general trend away from overt and institutionalized racism throughout much of the world. In the United States, for example, it is now socially unacceptable to openly judge a person as intellectually or morally inferior based on race alone. It still happens somewhere every day in America, but it's no longer mainstream or fashionable. For example, the public lynching of black men against a backdrop of joyous support from huge crowds and sanction from public officials was once a frequent occurrence in America but is now perhaps gone forever. Unfortunately, however, this fading away of overt and unashamedly racist acts in public does not mean that racism is gone. Now, in most cases, it's just more difficult to detect.

An experiment with disturbing results published in 2013 revealed that the mere proximity of a black hand or a white hand to a product for sale can significantly influence many potential buyers.[2] Researchers put iPods up for sale in more than a thousand online classified ads spread across the United States in some three hundred areas, incorporating both large cities and small towns. The ads included nearly identical photographs of a hand holding the iPod. But there was one key difference: in some photos, the hand was dark-skinned; in others, it was light-skinned. The ads with the black-hand image drew 13 percent fewer responses than the ads with white-hand images. They also received 18 percent fewer offers of payment to purchase, and those offers were 11 percent less, on average, than offers sent to ads with a white-hand image. Those people who responded to the black-hand ads also seemed less trusting than those who responded to the white-hand versions, according to the researchers. They were 17 percent less likely to share their names and 44 percent less likely to agree to a proposed delivery by mail (which would require them to share an address).

In another study, one with far more serious implications than

iPod sales, researchers asked people to shoot at armed criminals
when they appeared in a challenging video simulation that also
included innocent bystanders. Those shot most often—even when
unarmed—were black men.[3]

It is doubtful that these biased iPod shoppers and quick shooters
belong to hate groups and have swastikas tattooed across their
chests. I suspect that most of them probably think of themselves
as fair and nonracist. Probably few, if any, made the conscious decision to steer clear of black sellers or shoot unarmed black people in
a video experiment. More likely their shadow brain got the best of
them. It's a harsh reality to face, but no matter how wonderfully
enlightened you may think you are, your shadow brain might lead
you into unsavory behavior, too.

We all have troubling and potentially harmful biases against
people based on race, gender, age, nationality, religion, sexuality,
and/or other attributes and affiliations. Even those who belong to
groups that are the frequent target of prejudice tend to be biased
in favor of more socially valued groups. We shouldn't be surprised.
After all, blacks, gays, and women watch TV shows, films, and news
reports too. Their brains are subject to the same social inputs as
those within more favored groups. Subconscious racist acts—the
kind that occur without the oppressor's awareness—can be every
bit as destructive to people's lives as can conscious racism.[4] The
reason we may not believe ourselves capable of being biased is only
that these prejudicial views stay hidden away in our shadow brains.
They may emerge and speak to us when we meet someone for the
first time, poke us in the ribs when we hear an accent in someone's
voice, or tug at us when we are about to decide if a person is trustworthy, deserving of something, or dangerous.

Not only can the shadow brain pack an alarming load of negative
biases against other people based solely on their group membership,
but it may hold a few unflattering assumptions aimed at yourself as
well. And these personal negative biases can be detrimental to you
in many ways. For example, we may say and think that we are as
good or worthy as anyone else in some particular job or activity, but
hidden biases lurking in the shadow brain might surface and sabo-
tage us—without us ever being aware of it. Even the mere mention
of a negative bias can degrade performance in some people. This has
been demonstrated many times by social psychologists.

One study focused on a widely held belief that says women cannot do math as well as men. Researchers first tested this claim by giving men and women difficult math problems to solve. The men outperformed the women, which supported the stereotype. But when the researchers tested another group, they made one minor change: Before starting, they mentioned to the women that the test would not produce gender differences in the results. The women then performed as well as the males. A simple reassuring pre-test statement was enough to lower the "stereotype threat," which led to the women scoring higher.[5]

This is deeply troublesome, particularly when one considers the impact stereotype threat can have on children in school. Black students, for example, have been shown to underperform due to stereotype threat.[6] Even when black students were told that a test did not measure their ability, they performed worse when asked to check a box indicating their race before the test than when not asked to identify their race.[7] This problem is not limited to classrooms, of course. In one experiment, white participants were told that they were going to be given a test to determine "*natural* athletic ability," and as a result they practiced less before the test than did control groups who were not told that.[8] Presumably, they believed they were disadvantaged so they promptly gave themselves a disadvantage by reducing their commitment to preparation.

Don't be too hard on your shadow brain if you are starting to worry about just how unfair and prejudiced it may be. Being a renegade, generalizing, stereotyping, and jumping to conclusions without deep reflection is standard operating procedure for the subconscious brain. That's who it is; that's what it does. And, besides, it is natural and necessary. Our brain makes and takes shortcuts so that we can get stuff done. Most of the time it is harmless and helpful, but there are those moments when it results in significant problems for you or for others. Negative biases may come forth that both are factually wrong and contradict one's values and worldview. Truth, facts, and the ideals of civilized aspirations be damned, the shadow brain thrives on making sweeping generalizations because that's fast and simple. Sure, "you" can afford to spend minutes and hours reflecting on the deeper unity of humankind as you interact with others in daily life. Meanwhile, however, your shadow brain will take less than a second to feed you crude hunches about individuals based on superficial traits or markers and then it will suggest how you should interact with them. This is how sincerely fair-minded people end up doing things like not hiring the qualified minority candidate, paying female executives less than men for the same work, and deciding not to rent an apartment to someone with a "foreign" name. "They" may not be racists or prejudiced against gays, women, foreigners, and so on, but it hardly matters if their subconscious minds are. What makes this worse in some ways than legalized and other forms of overt racism is the lack of awareness that further hides it. A subconscious bias makes it possible for someone to feel completely innocent, rational, and fair, regardless of the harm he may be doing to another person by his actions or decisions. And, weirdly, maybe he is innocent. Are we responsible for everything that lurks in our shadow brains? If so, we all might be in trouble.

FERRETING OUT THE BIAS WITHIN

An important question to ask before going further is, how do we know there really are hidden biases against groups of people? How can we know if a good person has bad prejudices? We can know thanks to the IAT (Implicit Association Test)[9] and thousands of other related social-psychology experiments. The IAT, introduced in 1998, can expose subconscious biases against not only racial groups but also such biases as those against sexual orientation, age, skin color, body weight, height, disability, and nationality. The IAT has gone international, as tests are being given in thirty-nine countries in twenty-four languages with similar results. So far it seems like an effective way to shine a flashlight into our shadow brains. More than fourteen million people have taken one version or another of it, and the test site gets more than 20,000 visitors per week. The results have been both educational and deeply disturbing.

Participants who take the IAT are required to make rapid associations between words and images by striking a particular computer key. The speed of the test turns it into a battle between the reflexive, stereotype-filled, and fast subconscious brain and the contemplative, thoughtful, and slower conscious brain. With the luxury of time to think reduced, the raw subconscious is exposed. And it's often not pretty. For example, participants may be shown photographs of people with different skin colors and then asked to rapidly match them to words such as *hurt, awful, pleasure, good*. Those who are faster at making connections between positive words and white faces than for black faces, for example, have demonstrated a subconscious bias for white people and against black people.

Subconscious race biases can creep—or charge headlong—into many aspects of life, even those thought too important for such base emotions. It has been claimed, for example, that a person's racial bias won't affect an economic decision if it would be costly to that person. A 2013 study published in *Psychological Science* suggests otherwise, however. After monitoring economic decisions in relation to those with whom participants interacted, researchers found that "participants accepted more offers and lower offer amounts from white proposers than from black proposers, and that this pattern was accentuated for participants with higher implicit race bias. These findings indicate that participants are willing to discriminate against black proposers even at a cost to their own financial gain."[10]

Again, this problem is not limited to skin color. The IAT revealed that medical doctors, like the general public, tend to have a strong preference for thin people and a strong bias against overweight people.[11] How might this affect the way they treat patients? No good doctor would ever want to be unfair or neglect some patients due to their weight, of course. It is not difficult to imagine how it might happen, however, if the doctor is unaware of the hang-ups her or his shadow brain may have and does not make a conscious effort to ensure they don't play a role in professional decision making. "Cultural stereotypes may not be consciously endorsed," wrote the authors of a study on the negative impact these insidious biases have in healthcare, "but their mere existence influences how information about an individual is processed and leads to unintended biases in decision-making, so called 'implicit bias.' All of society is susceptible to these biases, including physicians."[12]

Researchers are careful to explain that test results indicating a significant implicit bias against one group or another does not necessarily mean that one deserves to be called a racist or a bigot. Don't forget that an implicit bias can be the opposite of what one consciously believes and in contradiction to how one behaves in society. However, as the researchers warn, these prejudices can sneak up on us and lead us to do things that are discriminatory against groups of people without us ever being aware of it.[13] Prejudice against someone based on some instant impulse of your subconscious is not only unfair to that person but is also bound to be unproductive for you many times throughout life. If you are hiring someone, buying something, or selling something, for example, you want to make the best possible rational decision, rather than some prejudicial, blind reaction. Therefore, it is in the best interest of everyone—and one more requirement for good thinking—to be aware of your known or possible implicit biases so that you can make a *conscious* effort to prevent prejudicial acts. Fortunately, it's not as difficult as you might imagine.

Encouraging research has shown that we can change or reduce our implicit biases in relatively easy ways. For example, simple exposure, being around or befriending people, can reduce a bias against the group they belong to. One study found that white people who had close friends who were black or Latino showed less implicit bias against those groups than did white people who had no Latino or black friends.[14] Awareness and determination can defeat or at least hold back these biases most of the time and in most situations. The shadow brain can be vulgar, ignorant, and dangerously assertive, but when people consciously decide to work at treating others fairly and judging them on individual merits, it is possible to quiet the bigot within. Mahzarin Banaji is a Harvard University experimental social psychologist and one of the founders of Project Implicit.[15] She believes the presence of implicit biases is not an insurmountable problem:

> We don't intend to discriminate or treat unfairly, but we do. Such unintended consequences come from aspects of our mind that seem natural (helping somebody close to us like a neighbor or a nephew rather than somebody more distant) and feels right (fearing somebody who looks physically different from us [or] strange). Such responses are natural and feel right because they evolved in a world where such responses may have been useful. And yet, they continue to operate even through the best person for the job isn't one's family member or friend, where in the strangeness of other cultures lie the most lucrative business opportunities. Becoming aware of the buggy aspects of our minds is the first step toward unraveling them.[16]

We can push back against this challenging source of fear, mistrust, and unfairness by actively seeking out images and data that prove the stereotypes untrue. For example, *male = strong* is a common subconscious bias that can hinder social progress and harm the efforts of women in business, school, and elsewhere. It can be weakened by viewing images of strong women and becoming more educated about women today and throughout history who have accomplished extraordinarily difficult things. Even simply *imagining* women in strong roles for a few moments before taking an IAT has been shown to reduce the subconscious *male = strong* and *women = weak* biases.[17]

You can take an Implicit Association Test online, through Harvard's website, and see if any subconscious biases against racial groups, overweight people, the elderly, gays, or women are creeping around inside your head. Don't be upset, however, if the results don't match up well with how you think of yourself. Keep in mind that the first step to fixing or improving a problem is identifying and acknowledging it. Lead IAT researchers Banaji and Anthony Greenwald offer good reason for hope, restrained though it may be: "We expect the next several years to produce a steady accumulation of research on methods to eradicate or outsmart mindbugs," they write in *Blind Spot*, their book about the IAT and subconscious biases.[18] They continue, "Although we presently lack optimism about prospects for fully eradicating mindbugs, we are not similarly pessimistic about prospects for research to develop and refine methods for outsmarting mindbugs."

I asked Banaji if it were possible that the IAT could reveal a negative bias for a group—African Americans, for example—not because the person has a subconscious dislike or fear of them but because the person understands and empathizes with the group's historical and contemporary struggles with injustice.

"We should be clear what we mean when we say 'implicit bias,'" she said.[19]

A specific test of implicit bias creates a measure of strength of association between concepts such as black/white and attributes such as good/bad, athlete/scholar, strong/weak, guilty/innocent. The sum total of all possible meanings of these attributes as contained in an individual's mind, at a given moment, is what the test detects. Implicit attitudes may be negative even in people who have sympathy for the history of black Americans—that's the nature of implicit attitudes and what makes them different from conscious attitudes, which are much more aligned with values about what's right and wrong. This is not an easy concept for most people to grasp, including some psychologists.

Banaji stresses that it is possible to keep one's subconscious biases in check with a bit of effort:

> Implicit attitudes as observed on current tests need not spill over into our behavior. My performance on the test may not change, but my behavior towards members of the social groups can change. That depends on the strength of my conscious attitudes and what I wish to do to modify them to change my behavior. If I am aware that I have a strong association of *good* with "young" rather than "elderly," I can ask where such attitudes come from, my daily experiences with young and elderly people, and my behavior. I can remind myself that as a university professor, my daily experiences include positive experiences with young people, that my knowledge of the elderly is deficient and that it is culturally mediated with strong input from the media. If I cared, I could choose to do things differently. I could choose to become a member of the AARP. I could decide that in any social setting in which the elderly are present, that I will approach rather than avoid, smile rather than frown or do nothing. I can make sure that I learn about the issues that the elderly face and give conscious thought to that which eludes me in my natural environment.

With all of this in mind, we can celebrate the fact that many traditional discriminatory behaviors are no longer legal or socially acceptable in many societies today and that increasing numbers of people say they hold the concept of fairness for all in high regard. But it would be both foolish and immoral to ignore the stealthy, more elusive sources of injustice that haunt our shadow brains. Our ability to be good thinkers, and good citizens of the world, depends on our acknowledgment of and active resistance to the negative underbelly of our shadow brain.

Chapter 9

THREE CRAZY THINGS THAT LIVE IN YOUR HEAD

"Life does not consist mainly—or even largely—of facts and happenings. It consists mainly of the storm of thoughts that is forever blowing through one's head."
—Mark Twain, *Autobiography*

W e have much to be mindful of during our waking hours. Numerous cognitive biases emerge from the shadow brain to guide our thinking and influence or direct our actions. Sometimes they help us; sometimes they harm us. It is easy to become intimidated, given the number of these inner forces and mental traps. Don't be. Simply do your best to learn about as many of them as you can manage. Understanding even a few is certainly better than knowing none. This chapter addresses three of the most common and problematic cognitive biases: (1) confirmation bias, (2) anchoring bias, and (3) hindsight bias. I consider awareness of these three, at least, to be necessary for good thinking. Watch out for them, in others and in yourself.

CONFIRMATION BIAS

When thinking about the many subconscious tricks and traps within our brains, there is no better place to begin than confirmation bias. This is the tendency we all have to lean toward everything that fits well with our beliefs and lean away from everything that challenges or contradicts our beliefs. It is one of the primary reasons why somewhat-reasonable people are able to hold totally unreasonable positions with great confidence. We spend hours, days, years noticing and remembering observations, evidence, and data that support a belief we already hold. At the same time, we fail to notice, or we promptly forget, observations, evidence, and data that threaten to expose the belief as false. This fuels everything from belief in conspiracy theories to political-party loyalties. It is impossible to sensibly estimate how much global waste and destruction comes from this one cognitive bias, but it must be immense. Everywhere, people walk around with smug conviction, assuming that they have rationally analyzed and judged the many claims and issues floating about inside their heads. In reality, however, their shadow brains have cheated by stacking the deck in favor of beliefs and positions, many of which may be highly nuanced, wrong, or just plain silly.

Confirmation bias is so pervasive, so natural, and so contrary to good thinking that it demands special attention. This one normal human cognitive process, perhaps more than any other, enables smart people to believe dumb things—and often with fierce and relentless resolve. It closes minds and supplies lost causes with an endless supply of ammunition so that we can fight on indefinitely. *Why should I doubt, reconsider, or give a fair hearing to an opposing view? I can't possibly be wrong because I know a million reasons why I'm right!* It drives people to continue defending a belief long past the point they should know better and give up on it. And none of us are safe. Highly intelligent people can be even more vulnerable to its consequences, for example, because their sharp minds might be more productive at selecting and storing data for their one-sided case. Confirmation bias has haunted humankind for a long time and probably isn't going away soon. Philosopher Francis Bacon wrote of it four hundred years ago:

> The human understanding when it has once adopted an opinion . . .
> draws all things else to support and agree with it. And though there
> be a greater number and weight of instances to be found on the other
> side, yet these it either neglects and despises, or else by some distinc-
> tion sets aside and rejects; in order that by this great and pernicious
> predetermination the authority of its former conclusions may remain
> inviolate. . . . And such is the way of all superstitions, whether in
> astrology, dreams, omens, divine judgments, or the like; wherein men,
> having a delight in such vanities, mark the events where they are
> fulfilled, but where they fail, although this happened much oftener,
> neglect and pass them by.[1]

It is crucial that readers understand how natural this process of selective observation, assessment, and retention of evidence is. One has to make the conscious effort to seek out, listen, and consider evidence that goes against the grain and feels wrong. It can be hard work and often is uncomfortable, but it must be done. Good thinking demands it. If you are a Christian, go to the trouble of actually listening when a Muslim tries to sell you on the Koran and Muhammad. It can be worth the time spent. If nothing else, you are likely to learn something new about a belief system you may not have known much about but is important to more than a billion of your fellow human beings. You may also learn things about your own belief system and your thought process about it as well. If you are an atheist, don't pretend that you have this and all other possible universes figured out from top to bottom. If they come calling, listen to the Mormon, the Scientologist, the Hindu, the Buddhist, and so on. Time is a finite resource, of course, so one wouldn't want to squander too much time on too many claims that may lead nowhere. But we all can find moments here and there to give alternative views a hearing. It's polite, it's mature, and it's good thinking. No matter how sure you are that you are right and they are wrong, *listen*. Keep in mind that we never know if confirmation bias has made a fool of us on a particular subject. It operates beneath our conscious awareness. So all we can do is stay humble and try our best to be fair to the contrary arguments and evidence we encounter.

Good thinking is a lost cause for those who never learn about confirmation bias, or who do but then fail to keep it in mind and act accordingly. People who do not take this seriously and push back against it are sure to pay the price of packing and hauling some or many unjustified beliefs for years. We all should think of it as the constant saboteur in our midst. "The confirmation bias is one of the most insidious and pervasive bits of software in your head," warns psychology professor Hank Davis.[2] "It is as much a part of being human as having two eyes, one nose, and two feet. To avoid evaluating the world through the confirmation bias, you have got to take conscious steps against it. Even then, there is no guarantee you'll succeed. If you allow your mental software to operate on its Pleistocene default settings, you will bring this bias into play."

California resident Chanel Prabatah, a thoughtful twenty-five-year-old with a degree in psychology, doesn't think the confirmation bias gets the best of her too often, though she does advise everyone to watch out for it. "I think people have ideas about various topics and end up confirming those ideas," she said.[3] "If someone says Kim Kardashian is a bimbo and has nothing to give to society, this person is probably going to then search consciously and subconsciously to find reasons why this is true. I feel like you have to be as objective as possible when you perceive life. But that can be hard because we are subjective humans who view life through different lenses because of our experiences. I honestly believe it takes a lot of conscious effort to avoid being manipulated by confirmation bias on a regular basis."

"Another reason to be aware of confirmation bias is to be able to see life for what it really is instead of trying so hard to connect dots," Prabatah continued. "Prime example: I recently bought some food for a friend. But when I phoned her she didn't respond. I began thinking of reasons why she wasn't answering my call, and then confirmation bias led me to put together a pattern in my head and start to think that she was mad and intentionally ignoring me. After a while, I told myself that I was going to give her food to a home-less person. But just as I was contemplating that, she called and explained that she had been in the gym working out. Confirmation bias almost made me give her food away for no reason."

Confirmation bias is one thing we cannot afford to be overcon-fident about. Good skeptics are no different from anyone else when it comes to the need to be vigilant about the confirmation bias. One cannot always be perfectly fair when assessing input, of course, but with effort it is possible to at least put up a meaningful fight for rational decision making most of the time. It's not easy. Once we make up our mind to believe or not believe something, it can feel like a waste of time to entertain another point of view even for a second. Trust me, I know. Once you have heard the arguments and "evidence" for a bad idea multiple times, it takes real dedication to patiently listen to one more attempt to convince you that this belief is the real thing. But we must whenever possible because of the principle at stake and the possibility that the argument we refuse to hear might have been the good one. No, I would not advise spending too much time listening to slight variations of the same argument over and over. However, when someone is presenting evi-dence and attempting to make a case, you need to consciously crank open the mind, give it a hearing, and restrain that confirmation bias. For example, after careful consideration, I came to the conclusion long ago that UFOs are probably not buzzing the planet on a regular basis and it's unlikely that Bigfoot populations are alive and well in North America. But these things could be true, and I never neglect to note this, whether to myself as a silent reminder or aloud to a true believer. We need to understand and never forget that the shadow brain is on the job every day, working hard at coddling us in the name of efficiency and psychological comfort. Confirmation bias is always there, stalking us, waiting for the chance to fill our heads with a lopsided case for something. Because I'm aware of how easy it is to be led astray by this bias, all of my conclusions, no matter how confident I feel about them, come with an escape clause. For example, I have attended and observed up close several faith-healing services, from the small-time to the big-time. As a result, I am pretty sure that faith healers are not tapping into magical powers to cure people. There are natural and much more likely explanations for what they do. But I don't pretend to know this with 100 percent certainty. There is always the chance that confirmation bias sabotaged my thinking. Maybe I have focused only on selective information and remembered only my observations that showed faith healers in a negative light. It's possible that there is good evidence somewhere out there that supports their extraordinary claims, but I have been so misdirected by my bias that I failed to find it. Awareness of the confirmation bias prevents good skeptics from clinging to and defending conclusions solely because they happen to be our conclusions. All knowledge is tentative, we say, pending the arrival of new and better evidence.

I once interviewed a woman who told me that she died, went to heaven (where she met Jesus), and then returned to Earth. She said Jesus had beautiful blue eyes, the streets were literally paved with gold, and there was a magical fruit basket in her room. When you took fruit from it, a new piece of fruit appeared.[4] Now, at no point during her story did I believe any of it was true—it was probably a case of confusion, delusion, and false memories, nothing too unusual for a human being to experience—but I still listened. You never know what someone is going to say until you hear them say it. Who knows? She might have pulled from her purse a magical, everlasting banana, a photo of heaven's gleaming roadways, a retinal scan of Jesus's blue eyes, or some other interesting evidence. She didn't, but I was there with open ears and mind if she had, ready to learn something new and adjust my worldview if necessary. On a regular basis I read articles, listen to radio programs, and watch television shows that push claims that oppose my tastes and worldviews. I routinely expose myself to what I consider the worst of political propaganda, pseudoscientific gibberish, and ravings of fanatical or dishonest religious leaders. Sometimes it is entertaining, but often it takes real effort to endure the madness. I do it, however, because I think it is important to hear ideas that are different from the ones currently encamped in my brain. It is contrary to good thinking to build a wall around oneself and avoid contact with opposing views. So what if they may be uncomfortable or seemingly insane? We don't want to hold on to positions simply because they already found their way into our heads or because they feel good. We want our conclusions to be as closely aligned with truth and reality as possible. And that requires keeping eyes, ears, and minds as open as possible.

If you are having any doubts about the importance of confirmation bias, consider that it may harm or even kill you, should you show up at a doctor's office or emergency room with a serious illness or injury. As human beings, medical doctors are under the influence of all the usual shadow-brain processes. Jerome Groopman, professor of Medicine at Harvard Medical School and author of *How Doctors Think* and *Your Medical Mind*, researched this impact of standard human biases on healthcare and has concluded that *most* medical errors are due to cognitive errors such as confirmation bias rather than technical errors. This means that the effectiveness of your treatment is less likely to be determined by *what your doctor knows* and more by *how your doctor thinks*. He advises patients to ask doctors simple questions about their condition such as, "What else could this be?" or "What is the worst thing it could be?"[5] This may nudge a doctor into rethinking a possibly flawed diagnosis.

In addition to confirmation bias, we need to be alert to a closely associated problem called *motivated reasoning*. Think of this as confirmation bias with a boost of nitrous oxide. Motivated reasoning goes beyond merely ignoring inconvenient facts and selectively remembering supporting data. This is about putting in extra, deliberate work to try to prove our beliefs while aggressively trying to destroy arguments against them. Turn on a cable-news channel any time of day and you are likely to see textbook-worthy examples of confirmation bias and motivated reasoning in action as politicians and pundits argue their positions. Truth and reality rarely factor into the rants and ramblings of these people. The potent mixture of confirmation bias and motivated reasoning over many years has left them well stocked with arguments for their conclusions and an unshakable confidence that they are always right.

Finally, never forget that confirmation bias and motivated reasoning can set up any of us at any time to wage passionate, lifelong wars in defense of bad ideas. Do not make the mistake of arrogance. You are not special when it comes to confirmation bias and motivated reasoning.

ANCHORS AWEIGH!

I must be careful about what I write to begin this section because it could have a profound effect on what you think the rest of the way. First impressions can steer us in unexpected ways, thanks to a subconscious trap called the *anchoring bias*. This is the tendency we have to latch onto the first available reference points, pieces of evidence, or input of any kind that we encounter, and then base further thinking and decisions on them.

The anchoring bias is similar to the framing effect addressed in chapter 7, but what makes this one even stranger and more problematic is that the initial anchor doesn't even have to be related to the topic in question. For example, crazy as it sounds, if I merely showed you the numbers 100,000 and 200,000 written on a piece of paper for a second or so and then a few minutes later asked you to estimate how many books there are in your local library, you would likely give me a significantly higher estimate than if I had first shown you the numbers 10 and 20. I know it seems absurd, but experiments have confirmed this effect again and again. Mere exposure to a number, a word, a fact, or an image can influence us because our shadow brain jumps on information without hesitation and utilizes it to help when we are faced with a challenging task. The shadow brain figures having something—anything—to go on is better than having nothing. The problem is that our conscious minds don't know this is happening, so we often end up working from a poor or nonsensical start point.

An interesting experiment by Amos Tversky and Daniel Kahneman produced a clear demonstration of the anchoring effect by asking people to estimate the percentage of countries in the United Nations that were African *after* spinning a roulette wheel.[6] Unknown to the participants, the wheel was rigged to stop on either the number 10 or the number 65. The roulette-wheel number was not supposed to have anything to do with the number of African nations. But it did in the subconscious minds of those who spun the wheel. The people who landed on 10 estimated on average *25 percent* of the United Nation members were African nations. Those who spun the wheel and landed on the number 65 estimated on average that *45 percent* were African. That's a meaningful difference generated by nothing more than the meaningless spinning of a wheel.

Another experiment by Tversky and Kahneman exposed how making calculations can influence our perceptions.[7] They gave the following math problem to a group of high-school students and asked them to come up with an answer within only five seconds: $8 \times 7 \times 6 \times 5 \times 4 \times 3 \times 2 \times 1$. That's not enough time to solve the problem, so the students had to make a guess after completing only the first few computations. The median estimate for this group was *2,250*. Tversky and Kahneman gave the following problem to another group of high-school students: $1 \times 2 \times 3 \times 4 \times 5 \times 6 \times 7 \times 8$. This group's median estimate was *512*. (The correct answer for the problem both ways is 40,320 because of the commutative property of multiplication.) There is a huge difference between 2,240 and 512, especially when you consider that it came about only because the test subjects did smaller or larger calculations before guessing.

Don't get the idea that this is all nothing more than abstract lab-room play. The anchoring bias directly impacts our lives. For example, it can play a key role in how we make purchases and negotiate with money. You may want to remember this when interacting with car dealers, real-estate agents, and human-resources people during salary negotiation. It can cost us significant amounts of money when our subconscious nudges us to make poorly reasoned decisions that are based off of arbitrary starting points that may lack a rational basis. Savvy politicians, salespeople, lawyers, and debaters know all about this bias, so you had better know too. If, for example, a political leader says, "We have calculated that saving our nation from the long, dark plague of bedbugs will cost the taxpayers more than $100 billion. I promise, however, that with my special cost-saving plan, I can achieve bedbug eradication for a cost of less than $50 billion." In this case, good thinking should take one's brain back to the original claim of $100 billion and question that figure before allowing the shadow brain to feel too agreeable and relieved that it will "only" cost $50 billion to end the scourge of bedbugs. Those who do not understand the anchoring bias are more likely to surrender to the lure and comfort of a reduced offer, even if it happens to still be unreasonably high. The anchoring bias is always there, ready to make us feel comfortable with overpriced items and unjustified ideas. Watch out for it.

GUESS WHAT ELSE MAY BE LIVING IN YOUR HEAD

To my knowledge, *Toxoplasma gondii* has not yet made its way into any science-fiction novels or horror films—but there is a surprisingly good chance that it's already in your brain.[8] This single-celled parasite, common in cats, may be infecting *two to four billion* humans right now. It wants to get into cats, its preferred host, so *Toxo* does weird things to the brains of rats that it infects like slow down their reaction time, reduce their fear of cats, and even make them intensely attracted to the smell of cat urine. All this makes infected rats more likely to be eaten by cats, which is good for the parasite. Scientists think people pick up *Toxo* from cat feces, cat scratches, undercooked meat, or contaminated fruits and vegetables.

Toxo may be manipulating human brains as well. Researchers say that infected men have slower reaction times and are less fearful, more reckless, and even more suspicious and jealous of romantic partners. Infected women may become more social and more trusting of strangers, which could be interpreted as a form of reckless behavior. One study found that pedestrians and automobile drivers in Prague who were involved in traffic accidents were more than twice as likely to be infected than other people in the same city.[9] That study has since been replicated. Rates of infection are significant. In London, 22 percent of pregnant women have tested positive for *Toxo*; in New York, 32 percent have; and in Paris, a shocking 84 percent of pregnant women tested positive for the parasite.[10]

Toxo may also be impacting or even causing serious mental-illness problems worldwide. Multiple studies have found that infected people are more likely than others to suffer from severe depression and anxiety problems. Some experts think *Toxo* could be responsible for one-third of all schizophrenia cases.[11]

Humankind has been coexisting with this parasite for thousands of years, so it's a little late to panic now. It won't help, anyway, as there is no current treatment to eliminate them from the brain once they are in. I recommend monitoring news about continuing research on *Toxoplasma gondii* in the coming years, as well as any other creatures that may be meddling with our brains. Our bodies are actually complex ecosystems in which bacteria and other microbes outnumber our own cells ten to one, so it should not surprise us to learn that some of these many life-forms that are on us and inside of us are capable of exerting an influence on our brains.

HINDSIGHT BIAS

It's easy to predict the future accurately after it happens. So easy that to do so would be silly, a waste of time, right? Absurd as it seems, however, we all do it. We look back at the past and perform a kind of confirmation bias in reverse to pick and choose just the right facts and feelings that make it seem that we knew what was going to happen, even though we didn't. When something important or unusual happens, we all have this natural tendency to tweak our recollection of the past so that we can believe that "we knew it all along." The shadow brain is a world-class architect of revisionist history. It is so dedicated to keeping us feeling smart, rational, and in control that it will feed its host a reconstructed view of the past that has been tailored and doctored to accommodate current emotional or circumstantial needs. It's also a way of ducking responsibility for something negative without having to consciously lie. *Hey, don't look at me. I knew the building would burn down one day. And I definitely warned somebody a long time ago about all those cans of kerosene stored next to the furnace.*

In the final moments of the 2015 Patriots–Seahawks Super

Bowl, Patriots defensive back Malcom Butler made a big-time play to stop the Seahawks' late drive and win the game. The unheralded and undrafted first-year player read the situation perfectly to get position over Seahawks receiver Ricardo Lockett and snatch the ball. Butler's theft of Russell Wilson's bullet pass from the one-yard line was a storybook ending for Patriots fans, but a climax that brought down a storm of emotional second-guessing from everyone else. Immediately, sports analysts and fans condemned the call by Seahawks head coach Pete Carroll. The consensus reaction, from casual fans to experts, was that Carroll's call was outrageously and obviously wrong. Why, they asked, would he not just give the ball to Marshawn Lynch, one of the NFL's toughest and most reliable short-yardage running backs? The Seahawks were on the one-yard line with twenty-six seconds on the clock and three chances to score. Why risk a pass? The only sensible call was a running play.

The reaction to the end of that Super Bowl was a textbook demonstration of the hindsight bias played out in a hundred million minds. *They blew it. I should have been coaching the Seahawks. I knew that letting Lynch run the ball was the right play—the only play!* After the game, Carroll defended himself by explaining that the personnel matchups were not good for a running play. This means the Patriots defensive players on the field at that moment were strong against the run and the Seahawks players on the field were more conducive to a pass play so, in Carroll's opinion, a passing play made the most sense. Moreover, quarterback Russell Wilson made a fine pass, and the receiver, Lockett, was in the right place at the right time. The only reason it didn't work out for the Seahawks is because Butler read the unfolding play perfectly and reacted to it exactly right. But what if Butler had been a millisecond too late getting to the ball or if he had dropped it? Then, you can be sure, very few (no one?) would remember thinking to themselves before the play that the only sensible call was a running play. More likely, millions of people would have remembered their thoughts in the final moments of the Super Bowl quite differently. *I knew that would work. The Patriots were expecting Lynch to run the ball, so I was thinking before the play that a pass was the perfect play to call. I knew it all along.*

Beyond Monday-morning quarterbacking, hindsight bias can have serious consequences. It can be a significant problem in many circumstances, especially when trying to figure out where things went wrong after a disaster. How can we understand a mistake and make sure it doesn't happen again, for example, if hindsight bias distorts our memories and analysis of the thoughts and actions that led to the problem? It is also a huge source of fuel for irrational belief. Answered prayers, astrological and psychic predictions, prophetic dreams, and visions can all seem remarkably accurate after the fact if the hindsight bias kicks in. If someone wins the lottery, for example, suddenly hindsight bias may have the person thinking about how those winning numbers had come to them over the course of previous days. *I saw them before I filled out my ticket. They were in phone numbers, on license tags in traffic, and my cable bill. It was meant to be.* No, more likely somebody was going to win and it just happened to be you. Rein in your hindsight bias and enjoy the money.

The hindsight bias can cause the biggest problems when people in positions of great power and influence are clueless about it. It can hamper a society's ability to learn from the past. This alone probably goes a long way in explaining why so many preventable large-scale mistakes, conflicts, wars, and disasters occur decade after decade. "If you feel like you knew it all along, it means you won't stop to examine why something really happened," says Neal Roese, a psychological scientist at the Kellogg School of Management at Northwestern University who has studied this problem.[12] "It's often hard to convince seasoned decision makers that they might fall prey to hindsight bias."

Hindsight bias is a problem in the legal system, too. Product liability and medical malpractice suits are particularly vulnerable to being sidetracked by it. A report by the American Association for Psychological Science states that it "routinely afflicts judgments about a defendant's past conduct."[13] Roese and other researchers believe that it is useful to consider opposite outcomes for an event and then think through the events leading to it. This kind of mental exercise can reduce hindsight bias, they say.

Back in the early 2000s, after Lance Armstrong won his third or fourth Tour de France, I wrote a commentary about him for the newspaper I worked for at the time. While acknowledging that everyone is innocent until proven guilty, I outlined several reasons to be suspicious of Armstrong's triumphs. I wrote that I felt it was virtually impossible for him to be dominating these brutally demanding races against the best riders in the world without doping because most, if not all of them, were using performance-enhancing drugs. I based this on private conversations I had with an experienced cyclist who had competed in Europe and the steady stream of reports about doping at the top levels of cycling. I reasoned that drugs were the most likely explanation for his extraordinary success.

Reaction to my commentary was fast and furious. Understand, this was during Armstrong's glory days when his popularity was at its peak. He was the cancer-beating, inspirational superhero everybody loved—except the French, of course. I was verbally pummeled, mocked, and condemned on local talk-radio shows. Angry Armstrong fans phoned me at work to complain. Several people demanded I read Armstrong's book, *It's Not about the Bike*, so that I might understand how wrong my suspicions about him were. Virtually no one I spoke with about Armstrong agreed with me back in the early 2000s. But a funny thing happened in 2013 when he showed up in Oprah Winfrey's confessional booth. After Armstrong admitted to using performance-enhancing drugs, virtually everyone I have spoken with about him claims to have never doubted that he had been cheating. They weren't fooled back then. They could see right through his lies. Of course he was doping, they say. I sometimes bring up Armstrong in casual conversations today just to observe the hindsight bias in action right before my eyes: "I knew it all along."

Sure you did.

Be alert to the hindsight bias in play during your life. Left unchecked, it will cloud your thinking and distort your thoughts about past failures and accomplishments. This matters because it will make it difficult or impossible for you to draw useful lessons that can serve you well in the future. I suspect that this is one of the reasons some people become trapped in cycles of unproductive or self-destructive behavior. They continue to repeat a negative pattern because hindsight bias prevents them from making a reality-based assessment of what went wrong in the past.

Chapter 10

AN ALTERNATE VIEW OF ALTERNATIVE MEDICINE

"Belief is the wound that knowledge heals."
—Ursula K. Le Guin, *The Telling*

*A*m *I dying? This is bad. What am I going to do? I'm going to
die out here.* Those were the thoughts I had over and over
during a long night in the African bush. Perhaps I was being overly
dramatic. But, then again, I had never felt so bad, never had such a
severe fever, and never before had been so sick so far from a doctor
or hospital. This happened while I was on a poor-man's safari in
Kenya. I was with a small group of strangers and a no-frills guide.
My temperature had gone from uncomfortable to possibly life-
threatening in a few hours. On top of my physical discomfort, I was
psychologically stressed, wondering if I had contracted yellow fever,
malaria, or worse.

Obviously I survived. But that experience forever influenced
the way I feel about unscientific medicine and the billions of people
around the world who believe in it. That long night in Africa certainly
could have led me to take something outside the bounds of medical
science. I'll never know, because there weren't any snake-oil sellers
or well-meaning-but-mistaken alternative-medicine practitioners
around to tempt me. Pain, fear, and desperation can degrade a person's ability to reason. I imagine those conditions can make virtually
any of us more receptive to foolish or fraudulent sales pitches. For
this reason I make an effort never to mock or belittle those who reach
for quack cures and dubious treatments in times of need. But polite-
ness doesn't mean silence. People need to hear the news that alternative medicine is not necessarily the safe and effective source of health
therapies its profiteers and cheerleaders say it is.

It would be wrong to condemn everything in alternative medicine, because it includes so much. This wild, no-holds-barred jungle

surely contains something good somewhere. A number of its products and treatments work to some degree, I'm sure. But if the scientific process is not the final word on what works and what doesn't work, what is safe and what is not safe, how are we to sort out the good from the bad? There is also the problem of what qualifies as alternative medicine. When someone asks me for my opinion on it, I first have to ask her what she means by "alternative medicine." Yoga, for example, is often categorized with alternative-medicine therapies. But who gets to define yoga? If a practitioner claims his yoga sessions will cure cancer or treat schizophrenia and, as a bonus, connect participants with their past lives, then, yes, it is properly labeled "alternative medicine" because there is no credible evidence to support those claims. But if a yoga session is billed as nothing more than stretching, strength poses, and relaxation, mixed with a bit of soothing music and socializing, then it likely has real physical and psychological benefits because a wealth of data indicate that such activities do. Meditation also is sometimes associated with alternative medicine but, as mentioned in chapter 4, there are sober indications that it may deliver a valuable boost to the mind and body. So some caution is warranted when criticizing such a wide-open field because blanket condemnations may taint something good along with the bad.

We also have to watch out for rejecting something good because uninformed or unscrupulous people made incorrect claims about it. Many nutritious foods have been hailed as miracle cancer preventatives and cures only to be exposed as having no evident influence on cancer rates—but they are still nutritious foods. I also readily admit the possibility that some apparently crazy and useless alternative-medicine products and treatments today might one day be shown to work. I'm also confident that some current mainstream medicines and treatments will be exposed as worthless in the future. Doubt and the readiness to change one's mind run both ways.

Paul A. Offit, chief of the division of infectious diseases at the Children's Hospital of Philadelphia, has taken on the anti-vaccination movement and written important books about healthcare. In his 2014 book, *Do You Believe in Magic? Vitamins, Supplements, and All Things Natural: A Look behind the Curtain,* he describes how he reviewed numerous alternative-medicine therapies to discover that most don't work—no surprise there—but also that many are dangerous. He offers this sampling:

Chiropractic manipulations have torn arteries, causing permanent paralysis; acupuncture needles have caused serious viral infections or ended up in lungs, livers, or hearts; dietary supplements have caused bleeding, psychosis, liver dysfunction, heart arrhythmias, seizures, and brain swelling; and some megavitamins have been found to actually increase the risk of cancer. My experience wasn't limited to reading medical journals. As head of the therapeutic standards committee at our hospital, I learned of one child who suffered severe pancreatitis after taking more than ninety different dietary supplements and another whose parents insisted on using an alternative cancer cure made from human urine. What I learned in all of this was that, although conventional therapies can be disappointing, alternative therapies shouldn't be given a free pass. I learned that all therapies should be held to the same high standard of proof; otherwise we'll continue to be hoodwinked by healers who ask us to believe in them rather than the science that fails to support their claims. And it'll happen when we're most vulnerable, most willing to spend whatever it takes for the promise of a cure.[1]

One of the key problems is that too many people have no appreciation for the meaningful border between science-based medicine and every other kind of medicine. "It used to be called 'fringe' or 'unconventional' medicine—or simply quackery," Offit explains. "Today, it's called 'alternative,' 'complementary,' 'holistic' or 'integrative.'"[2]

After working in a homeopathic hospital as a physician and then spending many years researching alternative medicine, Edzard Ernst concludes that one reason people keep lining up for it is because they are fed up with various aspects of science-based healthcare. They are angry that there are not better treatments for certain diseases. They are frustrated by their doctor's inability to spend enough time with them or by a perceived lack of compassion for their suffering. And then there are those who view alternative medicine as an opportunity to make one more fashion statement. "I think for many people it is a fashion thing, a sign of affluence to have all these useless treatments," Ernst told *New Scientist* magazine.[3] "Like the L'Oréal slogan, people think: 'I have reflexology or colonic irrigation because I'm worth it.' Then there is the image of CAM [complementary and alternative medicine] being 'natural' and that everything that is natural is safe, with no side effects. At the extreme end of the spectrum, people who are dying may be desperately searching for a cure. You would be amazed at the lies these people are sold. I don't hesitate to call it the criminal end of alternative medicine. . . . Integrative medicine is a subject that annoys me intensely. People are being told it is the best of both worlds—conventional and alternative—but when you look behind that platitude, it's a cover for quackery being smuggled into conventional use."

Ernst believes that the single biggest reason of all for alternative medicine's popularity is that many of the people who sell it are very good at lying. "It is a triumph of advertising over rationality: many of the forty million or so websites on alternative medicine promote outrageous lies. People seem quite gullible."[4]

Those who recognize the value of the scientific process applied to healthcare can only be alarmed at trends in the United States over the last few decades. One survey found that 42 percent of American hospitals now offer alternative-medicine therapies to patients.[5] "Wellness centers" selling some of the most absurd healthcare products and treatments imaginable are now found in strip malls everywhere. Many private doctors recommend untested or minimally tested treatments, supplements, and medicines to their patients. In 2011, *Consumer Reports* surveyed 45,601 of its subscribers to find out how many of them used alternative-medicine therapies. Three out of four surveyed said they used some form of alternative medicine for general health. But when *Consumer Reports* asked how satisfied they were with the outcome of these treatments, most were deemed "far less helpful than prescription medicine for most of the conditions."[6]

Use of alternative medicine may have doubled in the United States in recent years.[7] According to the National Institutes of Health, Americans now spend $34 billion a year on CAM (complementary and alternative medicine). Although it is clear that many people are wasting a lot of money and some of them are risking their health, don't get the idea that this has taken over America. Compared to the more than $2 trillion spent annually on conventional healthcare in the United States, this is still the minority player. But who can tell what the future holds? Moreover, America is not the only society with an appetite for alternative medicine.

The attraction to medical quackery and fraudulent healthcare products is a global problem, of course. Every country has too many unproven and disproven health remedies for sale. I have seen numerous blatantly unscientific health products and treatments in the Caribbean, South America, the Middle East, Africa,

Australia, and Europe. I once spent an hour or so exploring a little shop in Damascus, Syria, for example, that literally was stocked to the ceiling with powders and pills ground from every creature and plant imaginable. It was like an apothecary straight out of a Harry Potter book. Highlights included a couple of dried bats and the tail of . . . something, not sure what, hanging on a wall. At times I struggled to take in the dizzying sights around me. I found what may have been rhino horn for sale, as indicated by a picture of a rhinoceros on the jar. This was a disturbing reminder about one of the often-overlooked impacts of alternative medicine. Many threatened and endangered animals, including tigers and the Grévy's zebra, are killed for quackery.[8] The rhinoceros, one of the most beautiful and magnificent creatures on Earth, may soon vanish from the wild in part because of poachers who kill them for their horns. This happens because bad thinking allows quacks and crooks to profit off of gullible people by telling them the lie that ground-up rhino horn treats everything from a fever to devil possession. Knowing what I know about ethics and oversight in the alternative-medicine market, however, it would have been a good bet that no rhino horn was in that jar. There also was a staggeringly diverse assortment of herbal remedies for sale, far more than a typical western drugstore or supplement shop offers. As I scanned the shelves, the clerk quizzed me about any health problems or concerns I might have. He assured me that he had something for every known ailment. Of that I had no doubt. This is a business model that works everywhere because good thinking is not yet a standard feature of human societies. In Germany, for example, 75 percent of the population used alternative medicine at least once in 2010.[9] So much for that stereotype of Germans being more rational and logical than the rest of us. The United Kingdom is awash with alternative medicine as well, spending 4.5 billion pounds on it annually.[10] British writer Rose Shapiro, author of *Suckers: How Alternative Medicine Makes Fools of Us All*, writes:

> We are witnessing an epidemic of alternative medicine. There are as many as one thousand different alternative therapies, most with little in common bar one rather important thing: there's no evidence that they work. From chiropractic to color therapy, reflexology to Reiki such therapies are now used by one in three of us. . . . We [Brits] are increasingly likely to choose a personal consultation with one of nearly

fifty thousand alternative practitioners. These self-appointed and pre-dominantly unregulated therapists actually outnumber GPs [general practice doctors] in the UK. Now even the GPs are getting in on the CAM act—despite their scientific medical training more than half of British GPs now offer alternative medicine, either provided by them-selves or by referral.[11]

THE POWER OF NOTHING

Some of the unearned value awarded to alternative medicine is nothing more than the result of bad thinking and misperceptions about illness and natural recovery. People credit these products with helping them without knowing if it is deserved. What if, for example, I had consumed an herbal brew, sipped some homeopathic water, or been given a Reiki treatment that night in Africa? My body defeated whatever it was that took me down so hard. And it did so without any external help. I took no conventional medicine, not even Tylenol or aspirin. But had I turned to alternative medicine while battling the fever, I might have gone the rest of my days, telling the story of how some herbal concoction or homeopathic remedy saved my life. Many times I have had discussions with believers who share stories about how they were sick or had some ailment for which "regular" medicine or treatments did nothing. An alternative-medicine product healed them, however. My standard response is: "How do you know?" How exactly do you know for sure that it was the prayer or reflexology treatment that solved the health problem, when it might have been the antibiotics you were also taking at the time, or maybe it was the concurrent physical-therapy treatments? How can one know? Even if a person was taking an alternative-medicine product and nothing else, still it can be impossible or dif-ficult to know for certain that it was responsible for any improve-ment. After all, in most cases, it is time that does it for us. Human bodies are very good at fighting back against infection, healing injuries, and keeping us alive. They have to be because we are all under constant assault from the microbial world, and living entails many cuts, bumps, and bruises. I have had numerous illnesses and injuries over the course of my life, but none of them killed me. My survival rate has been 100 percent. That's right, to date, I have survived every minor and major health challenge—and so have you. You overcame every single one of them, probably the vast majority of them without medical aid of any kind. Time is the unsung hero of human health. Most people who contract the common cold and take one or more alternative-medicine treatments get better in a week or so. Most people who go to a doctor get better in a week or so. Most people who do nothing other than rest get better in a week or so. But nobody seems to want to give credit to time and the human body's defensive systems. Regarding my minor and major health challenges, I can say with certainty that alternative medicine had nothing to do with my perfect success rate—because I have never taken any kind of alternative medicine or treatment in my life. In most cases, my body got me through it on its own. In some of the more serious instances of illness or injury, over-the-counter science-based treatments or experts who rely on science-based diagnoses and treatments saw me through the storm.

The placebo effect has to be addressed in any discussion of alternative medicine, too. This bizarre phenomenon is real and surely accounts for much, if not most, of alternative medicine's continuing success and appeal. Still not fully understood, there is something about the power of expectation that can lead us to respond positively (or negatively in the case of a "nocebo") to medications and treatments that are fake, inert, or without an active ingredient. This has been studied extensively, and it can work in the context of both science-based healthcare and alternative medicine. If one hundred people with arthritis pain are given sugar pills, for example, and told by doctors that it is a new treatment that will ease their discomfort, some number of people in the group likely will report that they felt better after taking the fake medicine. Even sham surgeries generate the placebo effect. Patients have undergone fake operations and responded to them as if they had undergone real surgeries.[12]

Test subjects in some studies have also been found to experience the placebo effect even when told that the medicine they were taking was not real. But the effect is more likely to occur and be stronger if the doctor sells it well by emphasizing to the patient how great the medicine is, explaining how expensive it is, and/or generally playing up the complexity and importance of the doctor-patient encounter. Colorful pills work better than bland-looking pills.[13] Capsules are more potent than tablets, and fake injections stimulate the placebo effect better than do capsules. It also works better if the doctor wears a white lab coat rather than a regular shirt or dress.[14]

In light of what is known about the placebo effect, one begins to understand how all of those witch doctors and shamans managed to stay in business for thousands of years. It also helps to explain why some people praise pseudoscientific treatments such as therapeutic touch (manipulation of a patient's "energy" without touching the patient). The powers of our perceptions and expectations—even when wrong or based on nothing of substance—are extraordinary. Clearly the patient's shadow brain picks up on cues that trigger some sequence of events leading to, in some cases, an actual health benefit. Unfortunately, the placebo effect doesn't work all the time and doesn't work for everyone. And, making someone feel better doesn't necessarily mean anything positive has occurred regarding the actual health problem. This can lead to serious issues for people who interpret a placebo effect to be a cure. Imagine if someone takes an ineffective alternative medicine or science-based medicine, neither of which has any impact on the illness or injury. If the person senses improvement due to the placebo effect, she or he may figure everything is okay and, as a result, fail to seek treatment for a worsening condition until it is too late.

Good thinkers keep the placebo effect in mind when someone argues for an alternative drug or treatment. I recommend considering the placebo effect within science-based medicine as well. Even real medicine that works doesn't work for everyone, of course. It is not rude or unreasonable to question a doctor about your progress and ask if it could be a placebo effect you are experiencing. The standard question to ask someone who is encouraging you to take an alternative medicine is: Does it work better than placebo? If the person doesn't know or understand the question, then she or he is not qualified to be making healthcare recommendations to you. If the person says it is better than placebo, then ask for a source and check the study that is supposed to confirm it. Is it published in a reputable, peer-reviewed journal? Read the abstract, at least, and look for the researchers' conclusion. Is it firm or wishy-washy? How big was the sample size of the study? Generally, the more test subjects, the more reliable the study is. Be a good thinker. Do your homework.

SCIENCE IS NOT A DEMOCRACY

Another problem with belief in alternative medicine is that so much of its appeal and popularity are generated by nothing better than stories and personal recommendations. If someone says he took something and it made him feel better, that is considered validation enough for the typical alternative-medicine true believer. But a good story is not as good as a double-blind, placebo-controlled study—not even close. Good thinking requires one to be cautious and steadfast when facing a frontal assault from an appealing story. Resist. Think. Remember that our brains are capable of believing just about anything when delivered in the form of a tantalizing tale. Be prepared when your neighbor, coworker, or beloved family member tells you about the time an alternative-medicine product cured her of something. You can think through these moments and avoid placing your trust in something that at best doesn't work and at worst may harm you. We can't trust word of mouth when it comes to UFOs and the Loch Ness monster, so why would it be good enough for something as serious as your health?

WHAT IS A DOUBLE-BLIND, PLACEBO-CONTROLLED STUDY?

A basic example of a double-blind, placebo-controlled study would include two groups of patients of similar age and health status. Doctors or researchers would give one group a real medicine or treatment and give the second group a placebo, or fake medicine/treatment. (This could come in the form of a sugar pill. An example of a sham or placebo acupuncture treatment would be to touch patients on their backs with retractable needles that do not penetrate the skin.) The most important element of the study is that neither the patients nor the researchers know which group gets what. When done correctly, this is an effective way of eliminating intentional or unintentional bias from the patients and researchers so that the results are more likely to be accurate. In some experiments, a third group is included, one not given a placebo or real drug. This allows the opportunity to assess how much impact the placebo and the real drug had compared to people who took nothing. It's not perfect, but thousands of studies have demonstrated that it is the best way we have to figure out whether a drug or treatment actually has benefit beyond the placebo effect.

TWO WRONGS DON'T MAKE A RIGHT

Because defenders of alternative medicine almost always bring it up, it is necessary to address the imperfect nature of modern medical science. There is plenty of truth in the accusations alternative-medicine fans routinely hurl at science-based medicine. No doctor gets every diagnosis and treatment correct. Doctors make mistakes. Some doctors are incompetent. Some are unethical. Medicine that helps many patients may hurt others—possibly even kill them. There are numerous healthcare challenges and plenty remaining mysteries of the human brain and body that await solutions. Mental-health treatment, for example, is far behind other areas of healthcare. And, yes, some medical research seems far more concerned with profit than with saving human lives. But while all of this is true, none of it proves that a single alternative-medicine product or treatment works. Only the scientific process can do that.

Pointing to greed and failure in the medical-science industry does not make the greed and failure in the alternative-medicine industry go away. Science-based healthcare is highly regulated and tightly tied to scientific testing yet still has many problems. So why would this give anyone reason to feel better about alternative-medicine products and therapies that are mostly *unregulated* and loosely, if at all, tied to scientific testing? It makes little sense to run into the open arms of the alternative-medicine industry when it has all the same problems and more. Science-based healthcare will always be less than perfect so long as imperfect people do it. But even so, it is still immeasurably better than what we find out beyond its borders, where anything goes.

WHAT WE CAN LEARN FROM MEDICAL QUACKERY

Some critics view all of alternative medicine as nothing more than a vast swamp of lies and stupidity. I don't see it that way. I am convinced that modern science-based healthcare professionals can learn a lot from the people who make a living from homeopathy, chiropractic care, acupuncture, therapeutic touch, faith healing, and so on. First and foremost, they seem to understand the *human* aspect of healthcare far better than most medical doctors. Their stores, clinics, and offices tend to be warm and inviting. Look no farther than the waiting rooms. Calming music and incense fill the air. Attractive artwork on the walls. Living plants in the corners. The typical waiting room of a doctor with a science-based practice, however, will have a few old magazines, maybe an aquarium with a couple of sad goldfish, and that's it. One tends to be warm and hopeful; the other, cold and creepy. It's soothing vs. scary.

A year or so ago, my daughter had a minor problem and her regular pediatrician was away, so I took her to see someone else based on a referral. Within two seconds of walking into the new office, I sensed that something was off. The vibe was all wrong and my shadow brain went to red alert. The atmosphere was . . . way too nice and upbeat. *This can't be a doctor's office,* I thought to myself. It had the look and feel of a place one might actually want to go to and spend time. While I was wondering if we had wandered into a high-end jewelry store, the extraordinarily upbeat receptionist found my daughter's name in her appointments file, so we sat down. But I was still uneasy. I looked around, straining to make sense of my suspicions. And then I saw it: shelves in an adjacent room stacked high with herbal supplements. A dead giveaway. Of course, now it all made sense. The soothing music, pleasant smell, happy people, and likely quack products for sale on the side could only add up to one thing: we were inside a den of medical madness. I politely explained to the bubbly receptionist that something suddenly came up, grabbed my daughter, and escaped. Out of curiosity, later I looked up the place and, I was right, it was a "wellness center" specializing in holistic chiropractic, acupuncture, and reflexology.

I have explored many other alternative-medicine establishments over the years, not by accident but on purpose, and can't remember many that felt unwelcoming. I also cannot recall a single alternative-medicine practitioner I have interacted with who was not at least above average at talking and listening. One of the qualities many people cite in their defense of alternative medicine is that they are treated better in the customer-service sense. I don't doubt it. Alternative-medicine professionals seem to have more time for conversation than the typical doctor does. One can easily imagine how positive office environments and exceptional people skills work to maximize both the patient experience and the placebo effect. In fairness to science-based healthcare, most doctors don't rush their patients and neglect to ask relevant questions simply because they don't know any better or don't care. In most cases, they just don't have the time, because they are busy doing the important work of actually healing people and saving lives. But why not inject more humanity into the process?

Companies and governments invest billions of dollars into medical research every year in order to develop new drugs, invent new technology, and make discoveries. But some low-hanging fruit has been missed toward easing suffering and extending lives. It's possible, for example, that structuring healthcare systems to give doctors a little more time with patients and make simple improvements to the patient experience might significantly improve health outcomes. Adopting some of the customer-service techniques used so effectively by the alternative-medicine industry could do wonders. No one should have to feel like she is a car in for an oil change rather than a feeling and vulnerable human being in need of help. I suspect alternative medicine reins in and keeps many of its customers because its practitioners do a good job of listening to patients and making them feel like they matter. Imagine if healthcare systems everywhere made it a priority for doctors and nurses to give their patients more time and more attention. What would happen if science-based medicine stole the best pages out of alternative medicine's playbook? We might have a human-centered process coupled with medicines and treatments that work. The best of both worlds. How much suffering might that ease? How many more lives might that save?

A DOCTOR'S PRESCRIPTION FOR GOOD THINKING

John Byrne is a practicing internist and pediatrician in Michigan. He is also an assistant professor at Oakland University William Beaumont School of Medicine and the creator of Skepticalmedicine.com. Although aimed at doctors, the website is an excellent resource for anyone interested in the importance of critical thinking and skepticism in healthcare.[15]

Q. How serious do you think problems associated with alternative medicine/medical quackery are for America and the world right now?

A. People want alternatives. Americans spend more than $30 billion per year on complementary and alternative medicine [CAM] services and products. *More than $30 billion.* That is only one year in the United States alone. Clearly there is a big market for it.

That is the financial cost. There are also unquantified human costs. How can we quantify the costs of denying science in favor of implausible, unproven, or disproven treatments? We have a real problem when cancer patients (and parents of cancer patients) choose to stop chemotherapy in favor of unproven supplements or homeopathy as sold to them by CAM gurus. Denialism of AIDS leads people away from lifesaving treatments. Children with autism are being subjected to dangerous and implausible treatments such as chelation therapy, bleach enemas, hyperbaric oxygen, and chemical castration. Vaccine denialism is leading to the reemergence of deadly infectious diseases once thought eradicated from modern society.

There likely are many reasons that some people turn to CAM, including a sense of empowerment. The majority of those surveyed in the 1997–2005 *BMC Complementary and Alternative Medicine* report stated that they felt CAM empowered them to have a more active role in their own health. They claimed it emphasized treating the "whole person." These empowering notions are marketed very effectively. CAM gurus present their nostrums with authority and simplicity. They exploit notions of "vitalistic life force" and other forms of magical thinking. Believers feel empowered.

The survey also revealed a rising trend toward mistrust of mainstream medicine and fear of side effects. Many see mainstream medicine as being focused only on disease and not on people. These are problems that science-based medicine needs to address in order to keep people from turning to CAM.

However, as appealing as the false promises of CAM may be, there can be only one reality. The constancy of nature's laws is fundamental to our understanding of the world. Mainstream medicine has its problems for sure, but because it is informed by science, it self-corrects over time in light of new information.

CAM does not change in light of new evidence. It does not self-correct. CAM thrives on the misunderstanding of science and the promotion of magical thinking. Perhaps this is the greatest cost of all.

Q. What are your thoughts about the current trend toward "integrative medicine"? How do you feel about real doctors incorporating alt med into their practice and universities such as Harvard[16] getting involved with alternative medicine?

A. A nontrivial percentage of the population gravitates toward alternative medicine and is willing to spend a lot of money on it. This means there is an alternative-medicine market force that science-based institutions have to deal with. Big pharmaceutical companies know this and have become major producers of alternative-medicine "supplements."[17]

Healthcare systems also feel the stress of market forces. They compete aggressively for market share. They perceive that a significant portion of the public has a demand for services such as acupuncture, Reiki, and homeopathy. The administrators and physicians in charge often lack the critical-thinking skills needed to realize that such therapies have no scientific basis. Science is valued by these institutions, but the powers that be are often unaware of the fuzzy line between science and pseudoscience. Of the doctors who do realize the difference, many seem to shrug it off in a "whatever" fashion. Most of the regular medical staff at these institutions are likely unaware of what Reiki and homeopathy even are, let alone care enough to protest. Thus, our premier scientific healthcare institutions are bending to market demand and are promoting their new departments of "Integrative Medicine."

These departments purport to combine the best that conventional medicine and CAM have to offer. The marketing departments do a really good job at promoting such services to the public. Thus our academic institutions now find themselves profiting from pseudoscience. I fear that by integrating pseudoscience with real medicine, our scientific integrity has been damaged a bit, and most of us have not even noticed.

However, it is intellectually dishonest and confusing to promote science on the one hand, and sell pseudoscience on the other. It is kind of like allowing creationism to be taught as science in schools. Reality does not bend with popularity and market forces, but our ability to think critically apparently does. That is the real problem.

Q. As a doctor, what do you advise people to think to themselves or ask aloud when tempted to try an alternative-medicine product or treatment?

A. We should not expect most people to have the expertise to judge the scientific validity of any proposed treatments. The public has to rely on the authorities of scientists and science-based medical professionals. The problem is that most do not know the difference between legitimate authorities and fake authorities. People should be aware that legitimate science is an evolving, self-correcting process that represents the best information we have.

Science is not always right, but statistically it is the "least wrong" because it is always trying to prove itself wrong. Science is difficult to do correctly because it is human nature to do precisely the opposite—to try to prove oneself correct. That is why real scientific authorities have to prove themselves in our accredited institutions whose mission it is to ensure that the knowledge produced has survived the scientific process and the rigors of peer review. Real medical authorities realize that their opinions can never trump scientifically produced knowledge.

The public should learn the red flags of pseudoscience. Fake medical-authority figures will typically promote their own products and services without regard to the consensus of science. No amount of counterevidence will likely sway their recommendations. They will often use colorful terms when describing their nostrums, like "miracle" or "amazing." They may claim to have the "one cure for all disease." Pseudoscientists will be unable to actually explain their treatments, so they often hide behind words that seem almost magical, like "vibration" and "magnetism," "life-force," "energy," or more recently, "quantum."

When questioned critically, fake authorities may tip their hand by appealing to conspiracy theories with phrases like, "the cure that *they* don't want you to know about." They might make arguments from antiquity, claiming that their treatments have been used for centuries by ancient cultures. They often use the naturalistic fallacy by claiming that their product is "natural" and therefore is inherently good for you.

Fake authorities are not typically endorsed by accredited, academic institutions. Consumers should ask about university degrees and certification by recognized institutions such as the American Board of Medical Specialties. Dubious products are typically not endorsed by scientifically recognized authorities such as the Food and Drug Administration or the equivalent if outside of the United States.

When in doubt, the wise consumer should steer clear of products with the so-called Quack Miranda Warning: "These statements have not been evaluated by the Food and Drug Administration. This product is not intended to diagnose, treat, cure or prevent any disease."

WHAT IF THEY DON'T PUT THE MIRACLE IN THE MIRACLE CURE?

In 2015, New York Attorney General Eric T. Schneiderman announced that his office had conducted an investigation to look into the ingredients in popular herbal supplements being sold by major retail stores in his state. The results were appalling, though not at all surprising to anyone who follows the alternative-medicine industry.

Of the products tested, only 21 percent had DNA from the species of herb that was supposed to be present according to the label.[18] The other 79 percent either had no DNA from the correct herb or had

DNA from plant species that should not have been present. The products tested included: ginseng, ginkgo biloba, St. John's wort, and echinacea. Stores selling them included Walmart, GNC, Target, and Walgreens. The fillers investigators found included beans, rice, and houseplants.

There were, of course, angry comments tossed around in cyberspace in the wake of this announcement. People were upset because they had been defrauded. But, wait, what really happened here? New York's attorney general took action because products that probably do little or nothing for people did not contain ingredients that probably do little or nothing. What does it matter? What is the point of getting upset over whether or not an ingredient is present when no one has ever established that the ingredient works? It's like catching a conman selling nonexistent houses to people and then complaining about the absence of furniture inside those nonexistent houses. Echinacea, for example, was one of the herbal products tested that Walmart, GNC, Target, and Walgreens had on their shelves. The scandal is not that bottles of echinacea didn't have enough or any echinacea in them. No, the real scandal is that to date there is no body of credible studies demonstrating that echinacea works at preventing and treating colds, as the people who spend more than $130 million per year on it think that it does.[19] This is the state of the supplement industry today. Apparently it's okay to mislead or lie to consumers and sell them products that do not perform as suggested/promised—just so long as the product contains whatever unproven nonsense the label says it does.

One can almost sympathize with the CEOs of these supplement companies who cut corners and opt for cheap fillers. Why wouldn't they leave out an active ingredient that is, well, not really so active? I can imagine the big boss having an epiphany while addressing her board of directors: "Wait, why are we spending all this money to put real ingredients in our pills and powders? Since they don't really do anything, why don't we just replace them with dead houseplants and increase our profit margins? And we don't have to worry about this being unfair to our loyal customers because these products will still perform exactly as they always have. That's the real beauty of this idea—no one will notice! Can I get a show of hands? Who's with me on this?"

It is telling that genetic testing was needed to bring this problem to light. Why would that be? If these products were worth their high prices and actually performed as advertised, don't you think thousands of consumers would have been the first to notice that something was wrong and raise the alarm? Why was it left to investigators to discover that so many of these pills and powders are not what they are supposed to be? The answer, of course, is that the consumers who believe in this stuff are not likely to notice any difference between the correct herb and sawdust, because in most cases there is no meaningful difference in performance.

"Consumers already have ample reason to doubt most of the claims made by herbal supplement manufacturers, who have had precious little scientific evidence indicating these herbs' effectiveness in the first place," said nutritionist David Schardt in a statement released by the Center for Science in the Public Interest. "But when the advertised herbs aren't even in many of the pills, it's a sign that this poorly regulated industry is in desperate need of reform. Until then, and perhaps even after then, consumers should stop wasting their money in the herbal supplement aisle. . . . Attorney General Schneiderman has done what federal regulators should have done a long time ago."[20]

It is important to point out that this kind of investigation was not unique or new. In 2013, researchers in Canada checked the contents of forty-four popular supplements being sold by twelve companies.[21] They revealed that many of the herbs had been diluted or were missing. Manufacturers had used cheap fillers such as rice, soybeans, and wheat. In total, one-third of the supplements tested showed no trace of the herb named on the label. The scientists found that the echinacea samples, for example, contained *Parthenium hysterophorus*, also known as "bitter weed," which can cause nausea, rashes, and flatulence. One of the fillers found in the "memory enhancing" supplement called gingko biloba was black walnut. Ingesting this could have severe, perhaps even deadly consequences for someone with nut allergies. "Some of the contaminants we found pose serious health risks to consumers," the researchers wrote.[22]

Yet another study, this one published in 2012, found significant problems with black-cohosh supplements, marketed to treat menopausal symptoms. Researchers found that 25 percent of the thirty-six supplements they tested did not contain the correct species.

"Accidental misidentification and/or deliberate adulteration results in harvesting other related species that are then marketed as black cohosh," stated the study's authors.[23] "Some of these species are known to be toxic to humans."

In 2003 more researchers tested "Ayurvedic" remedies being sold in the Boston area and discovered that 20 percent of them contained potentially dangerous amounts of mercury, arsenic, and lead.[24] Two supplements were taken off the shelves in 2008 because they contained some two hundred times more selenium than their labels indicated. Consumers who had used them suffered muscle cramps, joint pain, blisters, diarrhea, and hair loss.[25]

In 2013, Paul A. Offit, chief of the division of infectious diseases at the Children's Hospital of Philadelphia, and Sarah Erush, clinical manager in the pharmacy department at the same hospital, co-wrote a commentary published by the *New York Times*, in which they warned consumers about problems associated with alternative-medicine supplements: "The F.D.A. estimates that approximately 50,000 adverse reactions to dietary supplements occur every year. And yet few consumers know this. . . . For too long, too many people have believed that dietary supplements can only help and never hurt. Increasingly, it's clear that this belief is a false one."[26]

To be fair, certain health supplements can be effective and may even be necessary for some people in some cases. We can't dismiss all supplements as fraudulent junk, of course. But why is this industry such a mess? Why can't one turn on a television these days without seeing an infomercial hawking suspicious or blatantly bogus health products? Mehmet Oz, "America's doctor," seems to have never met an alternative-medicine product he didn't like.[27] How can prominent retailers like Target, Walmart, and GNC continue to sell so many dubious products year after year? How did this happen? Why is an industry that earns $60 billion per year globally[28] allowed to be a free-for-all of lies and houseplant filler? Many people I speak with about alternative medicine are shocked when I explain that the US Food and Drug Administration maintains a mostly hands-off relationship with supplement manufacturers and sellers. They wrongly assumed that products used by millions of people, in many cases for very serious health conditions, are well regulated, tested, and monitored by the government. The reality is that companies have been left to police themselves, and the results have been precisely what any sensible person could have predicted. Based on investigative research such as I have cited here, rampant exaggerations in advertising, and the general lack of scientific support for products means that consumers are routinely cheated and many may be at risk. Short of people dropping dead en masse, however, the FDA looks the other way. It didn't have to be like this.

As with most things involving substantial amounts of money and sloppy thinking in America, there is a politician involved somewhere. In this case, that politician would be Senator Orrin Hatch of Utah. He was the force behind a 1994 law called the Dietary Supplement Health and Education Act. It has allowed companies to manufacture, market, and sell supplements without meaningful oversight.[29] The industry was essentially placed on the honor system. As a result, companies make health claims about their products—including everything from longer life spans to brain building to sexual virility—without having to back up any of it or submit products for government testing to determine effectiveness and safety prior to hitting store shelves. The *New York Times* reported in 2011 that Senator Hatch has consistently worked hard for the supplement industry, particularly in Utah, where several leading companies are based. Meanwhile, Hatch has received hundreds of thousands of dollars in campaign contributions from companies that benefit from the 1994 law.[30] It was also reported in the *New York Times* piece that his son, Scott Hatch, and at least five of the senator's former aides have worked as lobbyists for the supplement industry.

"Orrin Hatch certainly has a right to fight for his constituents," Steven Novella, a neurologist at the Yale School of medicine, told the *New York Times*.[31] "But the consequences are [that] we have an effectively unregulated market for these products, a Wild West, and people are being abused by slick marketing, and as a result taking things that are worthless or in some cases not even safe."

Every day around the world, people are harmed by alternative-medicine products and therapies—and sometimes they die. For example, a Harvard study published in 2008 found that more than 365,000 South Africans died prematurely from AIDS between 2000 and 2005 because the government promoted various alternative-medicine treatments including garlic, beetroot, and lemon juice over science-based antiretroviral medicines.[32]

If you consider yourself to be a smart and educated person, don't make the mistake of thinking that you are safe from exaggerated claims and anecdotes that push your emotional buttons. The alternative-medicine industry snares highly intelligent people every day. Chances are, people smarter than you have fallen for it. I have had numerous discussions about some of the worst medical quackery with bright and well-educated people who for one reason or another were staunch believers in it. This is not about grade point averages and professional accolades. This is about understanding how the brain can blind and betray any of us when we don't apply good thinking.

Apple cofounder and CEO Steve Jobs certainly was no intellectual lightweight, but he may have made fatal critical-thinking errors when he turned away from medical science and toward alternative medicine after being diagnosed with pancreatic cancer. According to one doctor, early surgery would have given Jobs a 95 percent chance of recovery.[33] Instead, Jobs chose to spend irretrievable time on acupuncture treatments, hydrotherapy, herbal supplements, bowel cleansing, and psychic healers. He died in 2011. Both Jobs's official biographer and a close friend said that Jobs had expressed regret during his final days. They said he recognized that he had made a mistake in trusting alternative medicine over medical science.[34]

Past experience informs me that this chapter will irritate and anger many people who are devout believers in alternative medicine. Because it is not my intention to upset anyone, I'll close with a conciliatory message for the believers: If I am a sensible person, I cannot condemn all alternative medicine, because I must admit that some of it surely is helpful. If you are a sensible person, you cannot support all alternative medicine, because you must admit that some of it surely is fraudulent and some of it surely is dangerous. As sensible people, we both support medicines and treatments that help and oppose those that harm or do nothing. This means we are on the same side. We want to be as safe and healthy as possible, and we want others to be as safe and healthy as possible. I am willing to accept, support, and use, if necessary, any medicine from any source—so long as the scientific process has shown it to be safe and effective. In this regard, the labels and categories we bestow mean nothing, ultimately. I hope that you will agree with me that the only thing that matters when it comes to life, health, and medicine is this: *Does it work?*

Chapter 11

GOOD THINKING VS. BAD IDEAS

"When men wish to construct or support a theory, how they torture facts into their service!"
—Charles Mackay, *Extraordinary Popular Delusions and the Madness of Crowds*

Bad ideas are as common as hopes and dreams. Wherever there are people there will be no shortage of twisted, lame, backward, and bizarre thought constructs that have little or no connection to reality. Maybe it has to be this way. Perhaps we need all these bad ideas so that we can recognize and appreciate the good ones by comparison. I don't know about you, but I'm a big fan of bad ideas. I can't resist paying attention to them, because they reveal so much about us. Most people shrug and dismiss Holocaust deniers, Loch Ness–monster believers, Area 51 enthusiasts, and the like. Not me. They are windows into the human brain's remarkable ability to believe just about anything. I advocate running toward these quirky mental contagions rather than away from them. The more of them we identify and analyze, the better we can be at good thinking. Somewhere inside every bad idea there is a valuable lesson to be learned.

Let's begin by considering a real conspiracy. It is widely known that John Wilkes Booth assassinated President Abraham Lincoln. But few people know much, if anything, about the *conspiracy* behind that terrible night of April 14, 1865. This is strange. What happened to the public's insatiable lust for conspiracy stories? This actual conspiracy involved a diverse and interesting cast of characters who plotted to kill Lincoln and other high-ranking officials in hopes of avenging the South by decapitating the federal government. But the general public knows little or nothing beyond John Wilkes Booth shooting Lincoln at Ford's Theatre. Why is this? Do

215

evidence, logic, and reality somehow make a conspiracy story less appealing? I have no doubt that if Booth had acted alone, there would be today a steady flow of dishonest "nonfiction" books and pseudodocumentaries claiming that there had been a conspiracy.

History and contemporary news overflow with similar *evidence-based* tales, many of them far more complex and unlikely than the Lincoln assassination conspiracy, but real events, nonetheless. Sadly, humanity has never been short of nefarious plans made in secret and put into action by groups of committed people. There simply is no need to dive into fantasy at the expense of good thinking. More examples from history:

- Concerned with his growing power, Roman senators conspire to kill Julius Caesar. They follow through with their plan and on March 14, 44 BCE, stab him to death in the Senate.
- A gathering of powerful Nazi elites takes place in a villa in suburban Berlin on January 20, 1942. They discuss and map out a course of action for the mass transport and murder of Jewish populations in German-occupied lands. This gathering of conspirators, known as the Wannsee Conference, put the horrifying reality to Adolf Hitler's bluster.
- From 1932 to 1972, medical researchers in what is known as the "Tuskegee experiment" monitor but do not treat nearly four hundred black men who have syphilis. They do this to study the long-term effects of the devastating disease—at the expense of the men's health and, in some cases, lives.
- In September 2001, nineteen men hijack four passenger jets and use them as guided missiles against targets in the United States. The al Qaeda conspiracy kills nearly 3,000 people.

Can't these true events satisfy whatever curiosity or lust for evil stories we may carry within? Shouldn't real killers in the shadows and real evil plots scare, entertain, and inform us better than imaginary ones? Yet so many people are compelled to proceed without evidence, sidestep logic, and give their trust to unlikely conspiracy theories. Why, for example, do millions of Americans believe that the US government was responsible for the 9/11 attacks when so much evidence points to al Qaeda? What is the seductive secret of conspiracy theories? How can many people claim to "know" that President John F. Kennedy was killed by a US government, US military, mafia, Soviet Union, or Cuban conspiracy (or all of the above!) when conclusive or strong evidence for any of those claims is absent? Having some doubt or suspicion about Lee Harvey Oswald acting alone is one thing, but claiming certain knowledge of a conspiracy is quite another. How is it that people around the world are so confident that the US government faked the Moon landings and have extraterrestrial bodies on ice or floating in vats of formaldehyde in some Area 51 bunker when no one has ever presented credible evidence for these things? Not trusting your government may be reasonable, but "knowing" that they are carving up aliens and reverse engineering interstellar vehicles is a delusion too far. Why are so many of us so vulnerable to this way of thinking? Once again, I suspect the answer lies not with insufficient formal education or an excess of stupidity and paranoia. No, the reason most people probably take these tall tales to be historical truths is because they have failed to account for the ways of that three-pound mystery machine they carry inside their heads.

"Belief in conspiracies seems to be an integral part of the human experience," said Nick Wynne, a US historian and former executive director of the Florida Historical Society.[1] "The interesting thing is that belief in conspiracies, like religion, is not dependent on the acquisition of provable facts. As a matter of fact, the absence of hard facts does not disprove a conspiracy, but, so it appears, is evidence that a conspiracy exists and the persons involved are powerful enough to cover up the facts. The less evidence a rational person can gather is proof positive, or so it would seem, that whatever conspiracy is being investigated is developed to such a point that it is able to assume a mantle of invisibility. Thus, no proof is the ultimate proof! The existence of 'conspiracies' is a matter of faith."

If people knew how much the human brain loves simple answers to complex problems, they might play hard to get when extraordinary and unlikely claims come calling. If they respected their shadow brain's ability to build a one-sided case for a claim, even as the conscious brain imagines it as being fair and considering all evidence equally, they might second-guess themselves and stop short of certainty when it comes to these wilder conspiracy claims. In addition to that, our thoughts and feelings about one thing often influence how likely we are to believe in another thing. People who did not trust or like President George W. Bush and Vice President Dick Cheney when they were in office, for example, are more likely to believe in a US government–led 9/11 conspiracy than are people who liked and supported those two political leaders. No atheists believe conspiracy claims involving the Antichrist because they lack the necessary preliminary beliefs or perspective for such ideas to take root in their minds. Few Christians lose sleep over *jinn* (genies) meddling in their daily affairs because the groundwork for belief in their existence was never laid. The tendency we all have to fall for bad ideas because they somehow coincide with or reinforce notions we already hold to be true is a significant problem that good thinking can catch. Be alert to "great" ideas or claims that happen to fit perfectly with your worldview. Maybe they seem great only because they align so well with your beliefs. And, like many such challenges, this can be easy to spot in others but difficult to identify in your own thinking.

Another reason conspiracy theories are able to grab so many minds and never let go is that they tease and challenge us. We love to figure out riddles, solve mysteries, and complete puzzles. Our brains typically reward us with a pleasant sensation when we do. Most popular conspiracy theories and other similarly amorphous and evidence-deficient ideas are little more than games people have mistaken for real life. This is a connect-the-dots puzzle with people and events as the dots. Don't forget our lust for patterns. Psychologists Daniel Simons and Christopher Chabris connected the dots between believing in unfounded conspiracies and seeing religious figures in food: "Conspiracy theories result from a pattern-perception mechanism gone awry—they are the cognitive version of the Virgin Mary Grilled Cheese."[2]

Michael Shermer, founding publisher of *Skeptic* magazine, has researched and written about many conspiracy theories and writes about them in his book, *The Believing Brain*:

> Why do people believe in conspiracy theories? I contend that it is because their pattern-detection filters are wide open, thereby letting in any and all patterns as real, with little to no screening of potential false patterns. Conspiracy theorists connect the dots of random events into meaningful patterns, and then infuse those patterns with inten-

tional agency. Add to those propensities the confirmation bias and the hindsight bias (in which we tailor after-the-fact explanations to what we already know happened), and we have the foundation for conspiratorial cognition.[3]

To understand more about the lure of conspiracy theories, we must understand something about the power that stories and gossip hold over us. First of all, we need to recognize that stories are virtually everywhere in human culture. We experience them in conversations, TV shows, films, songs, books, news and feature articles, poetry, and even in our dreams while sleeping. Interesting stories not only resonate with individuals, however, they are crucial to us as a kind of social glue as well. They are important in all societies "because they provide the framework that holds the community together," according to evolutionary psychologist Robin Dunbar of Oxford University.[4]

A typical conspiracy theory is essentially a negative story about a group of people stealing from, controlling, harming, or killing others. It is gossip magnified, and this explains much about their popularity. The story is the foremost learning and information-transfer mechanism we have. Our brains are predisposed to pay attention and remember stories above all other forms of communication. We can be sure that when language arrived, probably around a hundred thousand years ago, the story showed up right behind it. And the first lie soon after that. Or, perhaps, the first story was a lie. I wouldn't be surprised. During daytime walks to find food and especially while sitting around fires at night, undoubtedly, prehistoric people told stories about things that happened to them, things that happened to someone else, and things that never happened to anyone. We still see this behavior in traditional peoples today.

"There is something about fire in the middle of darkness that bonds, mellows, and also excites people. It's intimate," says Polly Wiessner, an anthropologist who has studied Kalahari tribespeople for forty years.[5] "Nighttime around a fire is universally time for bonding, for telling social information, for entertaining, for a lot of shared emotions." She calls stories "the original social media."

Telling stories—and our natural urge to listen intently to them—is only human. Conspiracy theories are nothing new. These negative stories, secrets about people, *juicy gossip*, are especially attractive to us because they are often interpreted as warnings. If some group of people is doing bad things, we may need to know about it for our own safety, so we listen and remember. The excitement of knowing a secret drives many of these bad ideas as well. Compare the Lincoln-assassination conspiracy with the John F. Kennedy assassination. The *real* Lincoln conspiracy story is no secret, yet nobody other than historians seems to care. But tens of millions of people know numerous details about *unsubstantiated* Kennedy conspiracy stories. We all have an urge to know things we are not supposed to know. Being aware of this urge can make you less likely to be sucked into the vortex of a bad idea.

Lisa Feldman Barrett, professor of psychology at Northeastern University, describes gossip as a "vital thread in human social interaction."[6] Gossiping is one of our uniquely human ways of sorting out who is trustworthy, dangerous, creepy, nice, and so on. Imagine how important gossip and conspiracy stories have been to us for so many thousands of years. If I am a prehistoric human and there is some weird guy over in the neighboring valley who routinely smears mastodon dung on his nude body and is rumored to have killed fourteen men with nothing more than his wits and a handheld live armadillo, I certainly would be eager to hear all about him. My brain would snap to attention and be more likely to remember such information if it came to me in the form of a good story. Don't just tell me to watch out for this freak. Tell me with building drama and plenty of detail how he gathers the dung before applying it. Is it fresh dung or old dung? How does he keep it out of his eyes? And what does the armadillo think about all this?

Gossip is so important to us that it can determine how and even if we see others. Barrett and her colleagues conducted an experiment in which they showed people photographs of faces that were tagged with negative, positive, or neutral gossip in a single sentence. For example, one came with the line: "Threw a chair at his classmate" (negative) and another was "Passed a man on the street" (neutral). Test subjects then viewed the face images without the gossip lines in a "binocular rivalry paradigm."[7] This means a face image was shown to one eye and a picture of a house was presented to the other eye, setting up a choice because the brain can only focus on input from one eye at a time. The participants were asked to strike specific keys on a keyboard to indicate which image they are looking at. Here's the interesting part: participants lingered on faces that had been associated with negative gossip longer than they did those associated with positive or neutral information. One brief exposure to a negative sentence made a face more interesting or important than others.

The researchers write,

> Gossip is a form of affective information about who is friend and who is foe. . . . Gossip does not impact only how a face is evaluated—it affects whether a face is seen in the first place. . . . These findings demonstrate that gossip, as a potent form of social affective learning, can influence vision in a completely top-down manner, independent of the basic structural features of a face.[8]

The researchers further point out that, while other primates bond through grooming, we prefer gossiping: "Instead of establishing and maintaining relationships by plucking fleas off of each other, we exchange and digest juicy tidbits of chit-chat, hearsay and rumor."

Remember to appreciate the power of a good story. Popular conspiracy theories and other extraordinary claims without sufficient evidence thrive because of their structure. They are tales tailor-made to exploit a weakness in our brains. Stories call to us. Gossip hooks us. We have used stories and gossip for many thousands of years to teach, learn, and warn one another about possible dangers. In fact, about two-thirds of all casual conversations are about people.[9] For all our lofty boasting about the magnificent power of language, it's mostly small talk that we engage in—and much of that is gossip with a negative slant. "If being human is all about talking, it's the tittle-tattle of life that makes the world go round, not the pearls of wisdom that fall from the lips of the Aristotles and the Einsteins," writes psychology professor Dunbar in his book, *Grooming, Gossip, and the Evolution of Language*. He continues, "We are social beings, and our world—no less than that of monkeys and apes—is cocooned in the interest and minutiae of everyday social life." It's easy to overlook the role of stories in our lives because we are always immersed in them. Fiction, stories mostly about people struggling or involved in some way with negative behavior, takes up two-thirds to three-fourths of the shelf space in most bookstores, with all nonfiction—from astronomy to zoology—squeezed into what's left. From my experience as a science and history teacher and giving talks about my books, I know firsthand the power of a story. More than once, I have come dangerously close to sending my audience into mass coma merely by sharing with them what I thought were exciting and important statistics. But then I watched their eyes come to life when I delivered information to them in the form of a short story.

SEVEN SIMPLE WAYS TO BEAT BAD IDEAS

1. **Know that your first thought may not be your best thought.** Remember that your shadow brain is probably going to make a snap judgement about a claim based on nothing more than a superficial assessment—an impulsive like or dislike—of the person or source presenting it. Your conscious mind has to acknowledge this and work to assess the evidence and arguments rationally. Being excited about a claim doesn't make it true.
2. **Use the embarrassment factor.** Try to imagine how dumb you will feel if a particular claim you are considering turns out to be wrong. This may seem silly, but it takes only a few seconds and it just might save you from making a few costly mistakes. Nobody likes feeling stupid, and we avoid it at all costs. Use this innate fear to your advantage. Let it motivate you to work harder when assessing the validity of a claim.
3. **Ask questions.** Good questions can expose bad ideas—even if you know little about the particular topic. The most basic questions are always appropriate. *How do you know? What is the evidence for this? Why don't the world's scientists/historians/experts agree with you?* Simply asking these kinds of questions can also be an effective way to get another person to think more deeply about the claim. Never let a belief set up camp inside your head for free. Make it earn its way in by the weight of good evidence and solid answers to your questions.
4. **Prepare your brain for battle.** Get enough sleep, eat well, keep learning, and stay physically active. A tired, underfed, understimulated, or sugar-soaked brain that is trapped in an unhealthy body can't do its best work. There are no guarantees, but a well-rested, healthy, and vibrant brain is at least in a good position to tackle life here on planet crazy.

5. **Recognize that there is not always a clear choice.** Sometimes we are confronted with claims that are deceptively framed as a choice between two options. *Do you want to become a Scientologist or do you want to live a diminished life burdened by anxieties and sadness?* Don't fall for this. There are almost always other possibilities.

6. **Admit you don't know everything.** There is the availability bias to consider. Our brains will reflexively grasp at whatever information is available, no matter how trivial, limited, or flawed it may be. Don't make important decisions or accept a claim because your shadow brain has made a fast and shallow read based on insufficient data. Sometimes we have to wait and take time to figure out what makes sense. It's okay to tell someone that you are not sure and will have to look into it and get back to him—even if you do have an immediate and strong feeling about it. No one can explain everything or come up with an answer to all questions. Sometimes a "miracle" can't be explained. So what? That doesn't prove that anything supernatural necessarily occurred. Ignorance is not evidence. We have to be grown-up, good thinkers and admit that we don't know what we don't know.

7. **Step outside of yourself.** Observe and listen. Notice how human you are. Appreciate that you have all the standard pitfalls, quirks, and vulnerabilities everyone else does. This can go a long way toward keeping you from stumbling into belief in bad ideas. Stay humble. Your brain has many important responsibilities, and it's doing the best it can. But when it comes to sorting out reality from fantasy, truth from lies, it's often overmatched. Help it by relying less on subconscious instinct and more on conscious research and analysis.

When we recognize conspiracy theories for what they are—gossip-infused stories—it is easy to see why they appeal to so many people and exist as a cultural phenomenon. There are deep subconscious and emotional factors at play when someone insists on telling you about the Illuminati or FEMA death camps. It's not really about facts and evidence, no matter what they say. This is something skeptics ought to keep in mind when trying to help someone think about these claims more clearly. Encourage these people to think about our natural tendency to latch on to gossip and stories that fit our particular worldview or fears. One should avoid the mistake of lumping all apparently baseless conspiracy claims together and automatically dismissing those who promote them. In all honesty, however, it can be a challenge to give some of them a full hearing. For example, in recent years, I have had not one but two people tell me independently on separate occasions that the Earth is hollow and there are huge, secret inhabited cities deep beneath our feet—and the government knows all about it but won't tell us. When I asked why the world's geologists haven't figured this out, the response from one was a shrug and the other was that geologists probably do know but won't or can't tell us.

I would guess that the reptilian-elite claim would make most top-ten lists of all-time bad ideas. If you haven't heard of it, don't worry. I researched this conspiracy theory so you don't have to. In brief, it involves a species of shape-shifting alien reptiles that have infiltrated the highest levels of human society and currently run the world. The claim has kicked around for years in various forms, but former British sports journalist David Icke gave it a popular boost in the 1990s with his book, *The Biggest Secret*. Prominent people such as Queen Elizabeth, George W. Bush, Hillary Clinton, as well as many famous actors and singers are supposed to be reptilian aliens. I readily admit that this would explain a lot, but, in all seriousness, no good evidence has been presented for any of it, so it can't be viewed as a rational belief at this time.

The numbers of people who believe in extreme and weird claims can be alarming. Most troubling is how these irrational beliefs might inform decisions and actions here in the real world. For example, a 2013 study found that *20 percent* of Republicans believe President Barack Obama is the Antichrist.[10] Yes, one out five Republicans in 2013 believed that the president of the United States is some kind of evil incarnate come straight from hell to unleash a diabolical plan on the world. Among independent voters, 13 percent say they believe this; and 6 percent of Democrats agree. Understand that these percentages translate to millions of adults in the United States. In 2015, former congresswoman, House Intelligence Committee member, and presidential candidate Michelle Bachman said that President Obama was accelerating the arrival of a supernatural apocalypse.[11] This person once wielded considerable power and influence at the highest levels of the world's most powerful nation—and she thinks a president's policies can bring about a magical end to civilization as we know it.

DEADLY DELUSIONS

Some conspiracy theories, although built on fantasy and little more, have the potential to do real damage. These have to be identified and confronted more sternly than others. Some people claim, for example, that there is a conspiracy by pharmaceutical companies, doctors, and governments to harm children with dangerous vaccines, all in the name of profit. They have no good evidence—only suspicions, hearsay, and cautionary tales about children with autism who also happened to have been vaccinated. Regardless, that is scary stuff to hear, and it can grab the attention of any parent. But those who make the claim that vaccines are harming millions of children are unable to make a reasonable case for it, so a parent's choice is between science, data, and expert opinion vs. non-experts telling scary stories about children who are autistic "because of a vaccine." Too many parents have turned away from the science to side with scary stories, and children suffer—and sometimes die—for it.

In 2010, a whooping-cough outbreak in California left ten children dead. Health officials linked the deaths to unvaccinated children. In the winter of 2014–2015, California experienced an outbreak of another easily preventable disease. More than one hundred people were diagnosed with measles, a highly contagious disease against which vaccines are very effective. Thanks to the huge success of vaccines in the twentieth century, many people today are unfamiliar with how serious measles, whooping cough, and other such preventable disease are. They seem to have no fear of illnesses that once terrified every parent. They are ignorant on the history of this matter, which makes them vulnerable to lies and pseudoscience. The reality is that these diseases are dangerous and deadly, and no child should suffer them if it can be avoided. The Centers for Disease Control describes measles, for example, as "the most deadly of all childhood rash/fever illnesses."[12] Nevertheless, one in ten adults in the United States now believes that the vaccine for it is unsafe.[13] This is occurring even as vaccines globally save more than nine million lives per year. UNICEF estimates that an additional sixteen million lives could be saved annually if vaccines were available to more people.[14]

Fearing and refusing vaccinations is not a new phenomenon, but this recent surge traces back to a 1998 article published in a prominent British medical journal that suggested vaccines cause autism. That article has since been retracted, and its lead author, Andrew Wakefield, was found guilty of "serious professional misconduct" by Britain's General Medical Council and was "struck off the medical register."[15] This was considered the harshest penalty, and it effectively ended his medical career in the United Kingdom. Siding with fraudulent research by a disgraced doctor against the scientific community's thundering evidence-based endorsement of vaccines is not good thinking.

We must also watch out for how runaway nonsense can impact politics. It seems in recent years that increasing numbers of American politicians are publicly embracing conspiracy theories, to the possible detriment of their country. I'm not suggesting that silly politicians with crazy ideas are anything new, but there does seem to be more of them in the last couple of decades. For example, more than a few elected representatives said, suggested, or hinted that President Obama was not what he claimed to be. They believed or fed the belief that he was in reality a Muslim Kenyan on a secret mission to destroy America from within.[16] Men and women elected to congressional and Senate seats have openly worried and warned the public about "Agenda 21," the alleged plan by the United Nations to take over America and strip away basic rights from all citizens.[17] It will usher in "new Dark Ages of pain and misery yet unknown to mankind," and "abolish golf courses, grazing pastures and paved roads," conspiracy believers warn.[18] Senator James Inhofe of Oklahoma said the following in a speech on the Senate floor: "Wake up, America. With all the hysteria, all the fear, all the phony science, could it be that manmade global warming is the greatest hoax ever perpetrated on the American people? I believe it is."[19] Not content to dispute the science on global warming, Inhofe claims that there is a massive conspiracy involving thousands of scientists, politicians, and government workers. Clearly, these days, claims once viewed as being rooted in nothing more than fringe paranoia can rise to the highest levels of power and take on an imposing sheen of respectability.

Do these politicians do this because they are weak skeptics and sincerely believe? Do they only pretend to embrace and promote these ideas as a way to manipulate voters? (Wait, did I just propose a conspiracy theory?) Who knows what is really going on inside their minds, but it can't be good for a society when people in positions of power and responsibility demonstrate such poor thinking skills. At any moment, more than half of all American adults believe in at least one discredited or unsubstantiated conspiracy theory. "These kinds of theories have the effect of completely distorting any rational discussion we can have in this country," Mark Potok, a senior fellow at the Southern Poverty Law Center, told *Newsweek* in 2014.[20] Many of these beliefs are tied closely to political ideology, but many are not and are shared by millions of people across political lines. Never allow yourself to think it is a problem that infects only one political party or specific worldviews. In the United States, it is true that minorities and the less-educated are more likely to believe in conspiracy theories, but large numbers of educated whites embrace them too. These are the conclusions of more than eight years of research by political scientists Eric Oliver and Thomas J. Wood.[21] They write:

> The brain did not evolve to process information about industrial economies, terrorism or medicine, but about survival in the wild. This includes a tendency to assume that unseen predators are lurking or that coincidental events are somehow related. Conspiracy theories reflect how we intuitively understand our world and, ironically, provide emotional reassurance. They are stories with good and bad guys, conflict, resolution and other narrative elements that have a natural appeal. In short, to adherents, conspiracy theories feel like the truth.[22]

"It is this that makes them problematic," Oliver and Wood continue. "By crystallizing intuitions into incontrovertible claims, they limit possibilities for public discourse. This might not be a problem if the conspiracy involves aliens. But when it comes to important issues such as gun control or vaccinations, conspiracy theories impede our ability to sustain public debate."

Oliver and Wood do not state that this pandemic of conspiracy-theory belief is a hopeless situation. But they offer no quick fixes either. They do, however, caution against turning every encounter with true believers into arguments over facts: "The first step should be to empathize. After all, whether knocking on wood or wishing someone luck, we all engage in magical thinking. Only by appreciating the emotional tug of conspiracy theories will it be possible for us to communicate in a meaningful way with our neighbors in tinfoil hats."[23] Sound advice because so little of this has to do with facts, reason, and logic.

When it comes to bad ideas such as popular conspiracy theories, we can do better and should at least try. Realistically, however, it's probably going to get worse before it gets better. The long reach of the Internet and power of social media, plus a steady flow of slick pseudo-documentaries on cable TV, make it a massive challenge to turn back the madness. My advice is to first make sure that your own good thinking keeps you safe from squandering time and energy on pointless conspiracy theories. *Then* worry about what others believe.

An effective way to reach believers, I have found, is simply to talk about *other* believers so it doesn't feel like a personal attack to them. I often explain, for example, that research has shown that those who believe in one conspiracy theory are more likely to believe in other conspiracy theories—even when they contradict. Really, it's true. People who believe that Princess Diana faked her own death are more likely than other people to believe that Princess Diana was murdered; likewise, people who believe Osama bin Laden was already dead when US Navy SEALs reportedly raided his compound in Pakistan are more likely than others to believe that he is still alive today.[24] Sharing information like this can help people recognize for themselves that many conspiracy beliefs are less about the claim making sense and perhaps have more to do with some general mistrust of government or a desire to explain events differently than the way they have been presented by official sources. And, of course, sometimes made-up answers feel better than real answers.

WHO CARES?

It is fair to ask if it is worth paying attention to bad ideas, conspiracy theories in particular. Does it make sense to spend time, energy, and ink addressing belief in reptilian celebrities? I think so. These beliefs may be silly on the surface, but scratch a bit and one finds real issues, significant ideas, fears, and problems. Conspiracy stories about the Moon landings, the JFK assassination, 9/11, Lady Di's death, and so on are but symptoms of a serious lack of trust millions of people have for the news media and their governments. And why do they feel this way? I suggest that much of it is well earned because schools, news media, and governments have shown themselves to be incompetent, unethical, and dishonest too many times. Perhaps the most dedicated conspiracy-theory believers are our canaries in the coal mine. They are millions of little warning lights telling us that our institutions and leaders need to do better. Perhaps we dismiss these believers at our own peril. In 1992, US historian Stephen E. Ambrose wrote an enlightening piece about the underlying weight of conspiracy theories in the *New York Times*. It included these lines:

> We should care because conspiracy theories about past events usually carry with them a political agenda for today. Erroneous or downright mythical views of the past can have important, even crucial, influence on the present. The coming to power of the Nazis, German rearmament, ultimately World War II might not have happened without widespread German belief in the stab-in-the-back conspiracy. Widespread acceptance by the American people of the "merchants of death" conspiracy thesis about our entry into World War I was a prelude to the ill-fated, nearly disastrous neutrality legislation of the 1930s. The unhappy consequences of McCarthyism would not have come about had the American people rejected his conspiracy theory about the triumph of Communism in China.[25]

Bad ideas, even the worst of conspiracy theories, matter. Remember those shape-shifting alien reptiles who run the world? Eleven percent of American adults either believe they are real or are not sure about it.[26] That equates to around twenty-seven million people. No matter how crazily convoluted their claims may be, we should never automatically dismiss conspiracy-theory believers as mentally ill or hopelessly dim. Most of them are only following a natural human impulse to absorb and share gossip. Some compassion is in order here because anyone can fall for a false story. Good thinking is the answer, but it does not just happen on its own, not given the way our brains work. One has to discover it, embrace it, and work at it every day.

Chapter 12

OKAY, SO NOW WHAT?

"Human existence may be simpler than we thought. There is no predestination, no unfathomed mystery of life. Demons and gods do not vie for our allegiance. Instead, we are self-made, independent, alone, and fragile, a biological species adapted to live in a biological world. What counts for long-term survival is intelligent self-understanding, based upon a greater independence of thought than that tolerated today even in our most advanced democratic societies."
—Edward O. Wilson, *The Meaning of Existence*

Let's imagine that people everywhere begin to hear whispers of an attractive and wise new way to live. They call it *Good Thinking*. Word spreads slowly at first among individuals and small groups here and there. Then it builds exponentially as millions learn about it and embrace it. Finally, critical mass is achieved and billions come aboard. Humanity as a whole finally marvels at the wonders of the strange but magnificent organ called the brain. For the first time, everyone questions everything. Second-guessing becomes second nature. In this unique leap of global human culture, reason is valued, evidence matters, and logical explanations are expected for all things important or unusual. Science is universally appreciated. Not worshipped, but understood and respected at last for its power to reveal reality and produce results. Voters in democracies become resistant to the shallow manipulations of politicians. They demand intelligence, competence, solutions, and progress from leaders and cast their votes accordingly. People living under the boot of dictators and theocracies may not physically escape their oppression, but their minds do. In this rapid, sweeping, and peaceful revolution, humanity transforms itself into a culture that prefers reason and reality to harmful traditions and distracting delusions. Societies invest resources only after careful, rational analysis. Projects and ideas that do not work are abandoned without reluctance. Lessons are learned, shared, and remembered. All ideas are welcome, but none are given a free pass. Evidence-based conclusions and creative possibilities rooted in reason become the foundation from which decisions small and large are made for both individuals and societies.

This near-universal good thinking does not usher in utopia or eliminate all of our destructive ways. We still struggle with many problems. But at least we do so standing fully upright for the first time, and without the heavy chains of superstition and self-inflicted blindness holding us back. At last we attempt to make the best use of the unique brain that more than three billion years of evolution bestowed upon us. Finally, we are the wise species, worthy of our name. *Homo sapiens sapiens.*

That would be an interesting world for sure. Certainly one worth trying on for size, don't you think? We shouldn't hold our breath waiting for it, however. Good thinking is the underdog, with a long, hard—some would say impossible—uphill climb to win over humankind. But one never knows. Stranger things have happened. After all, it wasn't so long ago that we were sea sponges and prosimians. Yet here we are today, listening to outer space in hopes of catching the murmurs of distant civilizations, visiting the ocean's deepest bottom, contemplating journeys beyond Pluto, and all the while making beautiful music.

Regardless of whether or not good thinking becomes fashionable for most or remains the fringe pursuit of a minority, science will continue its march. Things will change in the future. Things are changing *now*. Fred Gage, a leading scientist at the Salk Institute in La Jolla, California, told me that brain research is progressing today at a "faster pace than ever in the history of science."[1] This is a common perception among experts. "I think we're literally on the cusp of a complete revolution both in how we understand the brain and how we treat brain disorders," said Emad Eskandar, a neurosurgeon and associate professor at Harvard Medical School.[2]

What will we learn about the human brain in the next few years, the next few decades? How would society react to the unveiling of its remaining secrets? How will crime and law-enforcement change if new brain technology enables us to determine truth from lies, false memories from accurate memories? Imagine a society where no one can lie and everyone is a reliable eyewitness, one where criminal minds can be searched and seized *before* a criminal act is committed? Forget crime, what would total honesty mean for every marriage and friendship? And what about transhumanism—the enthusiastic coupling of new technologies with our bodies and brains?

What happens when the inevitable happens and there is a class of people with brains that have been turbocharged with expensive software, hardware, and/or drugs not available to all? How would the world's poor feel, knowing that they had been condemned to relative misery and mediocrity without escape because their parents could not afford genetic engineering or a neural upgrade for them? The current gap between the wealthy and the poor may seem trivial by comparison one day. Will civil wars and world wars be fought over a canyon between different kinds of brains? Or, what if the brain itself becomes the center of combat? It's far-fetched at the moment, but what if a government or company were to develop the ability to beam magnetic signals into brains to disrupt or control thoughts? Wars could be fought with fewer bullets and bombs but with far more cruelty. Brain wars aside, what would it mean for us if one day our heads contain the World Wide Web? Imagine something approaching the sum of all human knowledge instantly available within every mind. What if we could transfer our memories to an AI program that continues to learn and remember experiences from that point? Would this form of mental cloning qualify as immortality? What if we mix one mind with the minds of other people via new technologies? What might we be capable of creating? What solutions to complex problems might come within our reach, if we blend billions of brains into one massive mind?

We may not have answers today. But all of us certainly need to be thinking about these kinds of questions because many extraordinary possibilities are rushing toward us at increasing speed. We can be sure that science will continue to expose the brain's wonders at an ever-increasing pace in the coming years. And it is likely to be a far more important chapter in our story than any before it.

WHAT IF WE HAD A BRAIN REVOLUTION AND NO ONE CAME?

I recognize that I may be allowing hope and optimism to influence me, but I believe that something people will one day call "The Age of the Brain" is close. It may already have begun. Like many arenas of thought and innovation with profound implications, this one likely will be a mix of good and bad. Great discoveries often bring opportunities to exploit people and do harm. There is also the problem of apathy and willful ignorance. But even if the Age of the Brain were upon us and the majority of humankind were too busy with daily struggles or too distracted by nonsense to appreciate it, I believe that many people would care, enough to steer the world in a new direction. One individual with passionate feelings about the brain and good thinking can take up the slack for 10,000 who do not. I have spoken with many scientists who realize what may be near, and their excitement is infectious. This is a contagion I highly recommend exposing yourself to.

"The next one hundred years will be incredible!" said John Pfister, senior lecturer in the Department of Psychological and Brain Sciences at Dartmouth College.[3] "I think we will look back at fMRI [brain-scanning technology] and think about how crude or naive the technique really was. I sometimes ponder an odd neophrenology, where someone might come into a clinic, or even a therapist's office, and with a small, handheld device be able to stimulate an emotion, elicit a memory, or even scan for an abnormality with the wave of a hand."

Pfister admits significant problems and challenges may accompany the next steps and leaps in brain science but doesn't seem too worried about it:

> Did the discovery of the planets circling the sun turn society upside down? Did the discovery that germs cause illness turn society upside down? Did nuclear fission turn society upside down? Of course, of course, of course. [Progress in brain science] is neither good nor bad, but inevitable. The only thing that will be good or bad is what we do with the knowledge. I sometimes think that scientific advancement is like having children—hear me out. You ultimately can decide, for the most part, if you would like to take this step. The step itself is not trivial and should not be regarded as such. Once you have children, they are there, so they must be dealt with. Life with children is neither

better nor worse than life without children, but it is different. Life with children truly depends on how they are nurtured, understood, and ultimately allowed to thrive. The truth is that scientific advancement is inevitable, and that is how it should be. As a society, we cannot anticipate every change or even how knowledge will be used. But if the overarching principle is to treat each other with respect and remind ourselves that with knowledge comes responsibility, we should be okay. And, hey, who wouldn't want to live in a world that felt a lot like *Star Trek—The Next Generation*, to be exact?

Let's return to the present and all its concerns. It may be a profoundly difficult and even unlikely transformation, but no one can know whether it is *impossible* for humankind to become a significantly more rational and thoughtful species. Of course we never will escape the constant push and pull of the shadow brain. Of course we will always fall short of rational perfection. But with so much room for improvement, it's not difficult to imagine that optimistic *Star Trek* future Pfister speaks of being within our reach. And those who care about the welfare of others might consider the moral obligation here. For lack of good thinking, people are suffering and dying right now. If we can agree that the human brain is not only endlessly fascinating and immensely powerful but also profoundly strange and problematic in its ways, then our next step is to spread good thinking. We must tell. We must teach. We must show by example. People need to know. I was motivated to write this book for precisely this reason. Far as I could tell, no one I have ever had a ten-minute conversation with about the brain was bored. And credit goes to the subject matter, not my oratory skills. All who hear amazing facts, revelations, and stories about the brain are hooked and want to know more.

Perhaps you will make the decision to commit some of your time to promoting science, skepticism, and critical thinking to friends, family, and even strangers. I hope you will. It's not difficult. Just keep learning and keep sharing. Tell everyone you know about the brain, how it works, how it fools us, and what it requires to be healthy. They will listen. They will thank you for it. People heap gratitude on me, for example, virtually every time I talk about the brain's need for good nutrition, physical activity, and lifelong learning. Get on the right side of the Good Thinking movement. This is life-changing stuff. Who knows? Maybe it can be world-changing. If you feel that good thinking is good for humankind, then decide to be an ambassador for the human brain. Become a mental missionary, a freelance advocate for critical thinking. Maybe you can spark your own small revolution via social media in your corner of cyberspace. I know this much, we can't wait for more than seven billion people to figure it out on their own, one at a time. Those who understand the value of good thinking must share the news and share it often. How can we not? The brain is too valuable, too exciting for so many of us to ignore and misuse. This astonishing machine, this blob of endless wonder once led us out of a kind of prehistoric darkness. Now it can lead us out of modern darkness. In their book, *Shadows of Forgotten Ancestors*, Carl Sagan and Ann Druyan describe the brain as the tool we must finally decide to *pick up* and use with passion and purpose:

> If our greater intelligence is our hallmark as a species, then we should use it as all other species use their distinctive advantages—to help ensure that their offspring prosper and their heredity is passed on. . . . Our intelligence is imperfect, surely, and newly arisen; the ease by which it can be sweet-talked, overwhelmed, or subverted by other hardwired propensities—sometimes themselves disguised as the cool light of reason—is worrisome. But if intelligence is our only edge, we must learn to use it better, to sharpen it, to understand its limitations and deficiencies—to use it as cats use stealth, as walking sticks use camouflage, to make it the tool of our survival.[4]

Good thinking can improve lives now, and it has the potential to make the world a better place for everyone. But we have to recognize the difficult challenge of making it common. It is not the natural stance or behavior of our species. We evolved in a different time, under different conditions. Forget UFOs, we are the aliens here. This current, twenty-first-century environment, with all its social and technological complexity, is so new. But I believe (go ahead, call me an irrational believer) that we can change for the better, and not on evolution's unreasonably long (for us) timetable, but much faster. What we have to do is both make peace with and transcend the prehistoric brain we each are born with by exploiting its greatest quality, which is its flexibility. It can change, create, and adapt. It can solve difficult problems, including problems created by the brain itself. Our brains undergo constant physical changes, from months before birth until death. We can choose to take greater control of this process for our benefit. We don't have to be deluded by default. Good thinking is there for anyone to grab on to. If enough of us decide to change ourselves as individuals today, maybe our entire species can transform itself from believe-first sleepwalkers to reason-first thinkers. It is possible. But it won't happen on its own.

ESCAPE TO REALITY

Arizona resident Lane Kamp is a committed critical thinker who appreciates the value of science and skepticism.[5] He didn't always feel this way about good thinking, however. For more than three decades of his life, Kamp was a true believer, filled with unflinching faith and dedicated to a belief system built on extraordinary claims that many people would deem suspicious, if not outright nonsense. His emergence from it suggests that it is possible for us to break free from even the most personal and consuming irrational beliefs.

Q. How did you come to be a member of a religious organization led by a psychic/medium?

A. I never had a choice. My mother, Trina Kamp, claims she can leave her body, which allows a spirit teacher named Dr. Duran to speak through her. I was born into this nonsense. My mother traveled around the country with her husband, Steven, for many years in the 1970s and 1980s. They gave seminars and workshops on psychic development, rebirthing, and trance mediumship. People moved to Mesa, Arizona, to be near them, follow their teaching, and be close to Dr. Duran [the spirit]. This morphed into a cult around 1982. It was a small group that always fluctuated between 150 to 300 people. Many people came and left over a span of twenty-five years. Many families intermarried, and many children were born and indoctrinated.

Eventually my mother started a church called the Church of Immortal Consciousness. But the group always referred to itself as "The Community." Eventually they migrated to Tonto Village, Arizona. I would say my indoctrination started at about

the age of six or seven as I became more aware. I was indoctrinated via my parents' consistent daily preaching as well as
attending their weekly large-group meetings and trance sessions. I was involved with my mother's group until about the
age of thirty-five.

Q. What did it feel like to believe in the extraordinary claims
of "The Community"? Was it comforting? Did it make sense
to you?

A. It felt great to believe for most of my life. I was part of a
special group where everyone believed like I did. I was also in
a position of power, being the leader's oldest son, so my situation was a little unique. I had direct contact with the Holy
Spirit. Knowing you have a direct line to God and that you will
never really die is comforting. Yes, everything made sense until
I began to educate myself, read books, think more critically,
and ask questions.

Q. I understand that you were indoctrinated at a very young
age, but what enabled you to continue to believe these claims
as you matured? Did they appeal to your emotions? Was trust
in an authority figure a part of it? Were you afraid to doubt
and ask questions?

A. All of the above. I never questioned these claims until I was
much older. What son would want to suspect that his beloved
mother might be a charlatan? There was a huge emotional
appeal in the idea that life on Earth was short term and that
one day my family would all be together in some paradise. I
was very swept up by trust in my parents and not questioning
their authority or challenging their power. I was scared to be
outwardly skeptical, doubt, or question them because I didn't
want to be corrected, harassed, or bullied by my parents in
front of the group or by the other ministers.

Q. What changed in your mind that led you to figure out what
was likely going on and decide to leave?

A. I started noticing the corruption, manipulation, and abuse of power by my parents and the other leaders. They also started making requests of me and my wife and my children that shocked me into reality. One example was my parents' demand that my children marry whom they had chosen for them, namely other children in the group with very wealthy parents and grandparents. I also realized one day how arrogant the entire group was. To think that we had all the answers out of seven billion people on Earth and, out of all the other religions, our way was the one true way to God. That got me thinking.

Q. Do you agree that virtually anyone, regardless of intelligence and education, can be lured into a dangerous cult/religion/organization?

A. Yes! Anyone can fall victim. My parents' cult was filled with lawyers, teachers, people with master's degrees, psychologists, engineers, and several PhDs. If there was a mind vitamin called *skeptical thinking* and it worked, I'd chew a bottle a day.

Q. Many organizations tell us exactly what we want to hear in order to pull us in. What can we do to avoid falling for clever and seductive pitches?

A. I think understanding more about the sales-pitch process, how products and services are sold, can help improve skeptical-thinking skills. Storytelling, preaching, or pitches that sell ideas like gods, religion, magic, and pseudoscience products have the same story structure and format that's used in selling products and services that really do work. The Mormon religion and Scientology in particular are very good with their sales pitches. And [Texas megachurch preacher] Joel Osteen is too.

Q. What about people who are lonely, scared, sad, or feel their life lacks meaning?

A. If you are lonely, scared, or feel that life lacks meaning and it's getting bad, seek professional help from a licensed and experienced secular counselor or therapist. In my honest opinion, the worst thing you can do is take advice from your minister, family members, or friends. You need someone outside of your network, a one-on-one conversation with a professional.

Q. What advice can you offer someone who may already be in an organization that is based on dubious claims?

A. Stop and think for a moment about how arrogant it is to claim that your religion, your group, your teachings have all the answers. Educate yourself about what you really believe and the claims being made by the leaders. I think you'll start finding that things don't add up.

Q. What if a loved one is currently in such a group?

A. Keep the lines of communication open even if you have to lie. Seek professional help and get their advice on what to do and say. Whatever you do, do not send money to your child or family member.

Lane Kamp's childhood, teen years, and early adult life were spent deep inside a confining organization, trapped beneath many layers of irrational belief. His own mother was its leader. And yet he escaped. He doubted, asked questions, reasoned it out, and walked away. If he could manage that, then others caught up in far less oppressive irrational beliefs are likely capable of breaking free as well. The first step is good thinking.

People are ready to hear about a new and better way. They are always ready. Look around, bad ideas are never short of devotees. Surely a great idea—"Know, use, and care for your brain well"—can gain traction. People everywhere are still looking for wisdom and crying out for help. This is why they eagerly follow so many barely camou-flaged lies and bad ideas. No culture, no nation, no government or belief system, certainly no one person has it all figured out. Solutions to problems, both personal and global, are promised every moment. Yet we are always short of solutions. Good thinking is not the answer, either. It is the crucial step that may lead us to the answers. All people in all societies can benefit from hearing the strange truth that their brains can be their best or worst possession. It can lift them up now or hold them down forever, depending on how they choose to use it. Presented well, with kindness and sincerity, the call for good thinking might be an easier sell than expected because ultimately this is a simple thing. It is encouraging and promoting human potential to humans.

What would happen if most people aimed their mental energies toward good thinking? What if everyone knew that our eyes and ears lie? Imagine if everyone understood that we are mostly passive passengers driven around by a shadow brain? How might good people change if they realized that our brains take instant shortcuts behind our conscious backs, many of which are not logical, helpful, or fair to others? What if virtually everyone knew that good thinking is the only good defense against dangerous and time-wasting delusions and resolved to put more thought into their thoughts as a result? I welcome correction, but I can't think of anything that might be more important than promoting critical thinking, skepticism, and basic brain education for all. So many of our most serious global problems flow directly from the shortage of good thinking. Poverty, the loss of biodiversity, malnutrition, healthcare, sexism, racism, war, terrorism, and so on. Good thinking deserves a place on the short list of humankind's basic necessities, right up there with security, healthcare, and education. Of these, good thinking would be the least expensive to achieve, by the way. Perhaps I am over-reaching—I'll leave that to you to think about—but good thinking may well be the only thing that can remedy our increasing risk of self-destruction.

And clearly we do need saving. In 2015, the Islamic State, or ISIS, worked hard to make the world cringe in fear and disgust by releasing videos of mass shootings and choreographed beheadings. The worst may have been its video of a captured Jordanian pilot burned to death inside a cage. The violent behavior of these people disturbed me most of all, of course, but I was also troubled by reactions from the public, pundits, and politicians in the West. Virtually every voice I heard in the wake of every ISIS action failed to identify the problem they sought to solve. Most liberals, for example, talked about thugs, psychopaths, and gangsters "perverting Islam," while most conservatives declared that Islam was the problem. Some analysts claimed more nuanced causes were to blame. These included nationalism, poverty and unemployment, revenge for Western foreign policies and wars, and so on. All, however, missed identifying the real foundation that this madness and all else like it rests upon.

The Islamic State, the particular group that rose up to create a new spasm of violence and chaos in the Middle East are not *the* problem. They are only the latest *expression* of the problem. They are a symptom of a disease that has been making us sick for a long, long time. They are but smoke from the fire that is bad thinking. Members of the Islamic State, for all their different origins and backgrounds, have one thing in common: They aren't thinking well. For example, people who have committed themselves to skepticism and critical thinking would never follow orders or take guidance from religious leaders who say the world is 6,000 years old, life does not evolve, no one ever walked on the Moon, genies are real, and a supernatural end of the world is going to happen very soon. A good thinker wouldn't cross the street on the command of such a misinformed and deluded person, yet thousands fight, kill, and die on the word of such leaders. Islamic State fighters are told they are destined to defeat the "army of Rome" (the West or Christians in general, perhaps) at Dabiq, Syria, very soon. They believe that this conquest will then trigger the final stage of the countdown to a supernatural apocalypse that will climax with the return of Jesus, who will lead the Islamic State to ultimate victory on planet Earth.[6] Yes, Jesus.

Journalist Graeme Wood provided an articulate roundup of the bad ideas that fuel ISIS in a 2015 article published in the *Atlantic*: "The Islamic State . . . follows a distinctive variety of Islam whose beliefs about the path to the Day of Judgment matter to its strategy. . . . Its rise to power is less like the triumph of the Muslim Brotherhood in Egypt (a group whose leaders the Islamic State considers apostates) than like the realization of a dystopian alternate reality in which David Koresh or Jim Jones survived to wield absolute power over not just a few hundred people, but some 8 million."[7]

We can assume there are some psychopaths and more than a few thrill-seeking young males among the ranks of ISIS, but certainly many, probably most, of these people are fighting for a cause they believe makes sense. Good thinking could have kept them at home. Good thinking would have compelled them to doubt and ask for evidence the first time someone tried to convince them that a magical end to the world is close at hand. They never learned or they failed to appreciate the vulnerabilities and deceptive ways of the shadow brain. They didn't place enough value on evidence, science, and skepticism. So they ended up playing the role of murderous puppets for a senseless cause that only brings more hate and misery to the world.

Bullets and bombs may or may not provide momentary relief to problems such as the one the Islamic State and other groups like it pose to civilization. But it is only good thinking that can address the real source of such behavior and offer the possibility of lasting relief. In the end, it doesn't matter what religious or political label such groups carry. The problem is not Muslim terrorists. The problem is that sane people can be led to insane behavior when good thinking does not protect them.

We don't have to look back centuries to the Crusades for examples of similar unlikely claims fueling extreme violence and propping up deranged Christian armies. Ugandan warlord Joseph Kony led his "Lord's Resistance Army" on a reign of extreme violence through Central Africa that has slowed in recent years but still is not completely over. This former Catholic altar boy claims to be led by the Holy Spirit on a mission to establish a Christian government based on the Ten Commandments. His soldiers—infamous for torturing and murdering civilians, raping women, and using children as sex slaves and fighters—are required to paint a Christian cross with holy oil on their chests before battle. Any soldier who neglects to do this is subject to execution per Kony's orders.[8]

Just as many do regarding the Islamic State, one can argue that Kony has used Christianity as an excuse or distraction and that he is not a "real believer." Maybe that's true. One can never be sure what goes on inside the minds of politicians and warlords. But surely many of Kony's thousands of followers over the last few decades have sincerely believed that they were killing for Jesus. I see no reason to doubt this, because I know that in the absence of good thinking people are more than capable of believing anything they are told by authority figures and encouraged to accept by popular tradition.

What matters most in all of this is that too many people are willing to harm themselves, others, and the entire world for poor reasons or for no reason. The Islamic State, the Lord's Resistance Army, and all the groups that will follow them in similar quests should be seen for what they are: new eruptions of the same underlying constant failure that has burdened humankind for thousands of years. This is bigger than individual armies and religions. Focusing our attention on the new alliance of the month or the latest label to fear is a waste of time because it obscures our view of the real challenge. It's like fixating on the color of a house while it is burning down. The color is irrelevant. It is the flames that matter. Good thinking addresses the flames. It puts out fires. Even better, it can prevent them from starting in the first place.

Only a new relationship with our brains and the spread of good thinking across all cultures offer us a possible escape from the constant cycle of delusion and self-destruction. Most of the people bad thinking harms are far removed from warlords and terrorists. What do we do about them? Violence certainly isn't the answer. We can't use bullets against the elderly woman who squanders her money on medical quackery. We can't drop bombs on the man who votes for the demagogue promising to insert creationism into science textbooks. Both on and away from the battlefield, our real struggle is against ignorance and delusion—not so much against fellow humans as against bad thinking in all forms. This continual crisis can be overcome only by elevating reason because a drought of reason is the actual crisis. Similar to the way Gandhi and Martin Luther King figured out how to turn millions of hearts from hate to acceptance, we must work out how to shift billions of minds from irrational belief to reason, from bad thinking to good. We need to develop an international formula for reaching people so that they can appreciate and desire the freedom, joy, and benefits that good thinking offers. Whatever that formula may be, it begins with you. By leading a life rooted in reason, you can show others the benefits. By explaining the brain's wondrous, surprising, and challenging processes, you can help change the world one mind at a time.

Encouraging good thinking is a relatively easy undertaking but one with profound potential impact. We cannot impose it or give it to others, however. All we can do is point at it, explain it, demonstrate it, and hope that they decide to pick it up. This dream of a more rational and thoughtful species might well be impossible to achieve. Maybe our brains are just too wild to corral and our cultures too saturated with irrational belief to save. I won't pretend to know that it can be done. But I certainly do believe that good thinking is a summit worth reaching for.

NOTES

CHAPTER 1: THE CASE FOR GOOD THINKING

1. Emily Dickinson, *The Complete Poems of Emily Dickinson* (New York: Back Bay Books/Hachette, 1976), p. 312.

2. Kenneth Feder, interview with the author, February 27, 2015.

3. Joyce Ehrlinger, Thomas Gilovich, and Lee Ross, "Peering into the Bias Blind Spot: People's Assessments of Bias in Themselves and Others," *Personality and Social Psychology Bulletin* 31, no. 5 (May 2005): 1–13, doi:10.1177/0146167204271570.

4. Paul Offit, *Do You Believe in Magic? Vitamins, Supplements, and All Things Natural: A Look behind the Curtain* (New York: Harper Paperbacks, 2014), p. 42.

5. Cameron M. Smith, interview with the author, February 11, 2015.

6. John F. Pfister, interview with the author, February 22, 2015.

7. John Byrne, interview with the author, February 18, 2015.

8. David W. Moore, "Three in Four Americans Believe in Paranormal," Gallup, June 16, 2005, http://www.gallup.com/poll/16915/three-four-americans-believe-paranormal.aspx (accessed June 16, 2015).

9. Linda Lyons, "Paranormal Beliefs Come (Super) Naturally to Some," Gallup News Service, November 1, 2005, http://www.gallup.com/poll/19558/paranormal-beliefs-come-supernaturally-some.aspx (accessed January 2, 2015).

10. Ibid.

11. Pfister, interview.

12. Fred Gage, interview with the author, June 17, 2015.

13. Smith, interview.

14. Jerome Lawrence and Robert E. Lee, *Inherit the Wind* (New York: Ballantine Books, 2003), p. 93.

15. Feder, interview.

16. Harris Poll, "What People Do and Do Not Believe In," Harris Interactive, December 15, 2009, http://www.harrisinteractive.com/vault/Harris_Poll_2009_12_15.pdf (accessed January 2, 2015).

17. Pew Forum on Religion and Public Life, "Many Americans Mix Multiple Faiths," Pew Forum, December 9, 2009, http://pewforum.org/Other-Beliefs-and-Practices/Many-Americans-Mix-Multiple-Faiths.aspx (accessed December 22, 2014).

18. Quoted in Tom Head, ed., *Conversations with Carl Sagan* (Jackson: University Press of Mississippi, 2006), p. 120.

19. Frank Newport, "In U.S., 42% Believe Creationist View of Human Origins," Gallup, June 2, 2014, http://www.gallup.com/poll/170822/believe -creationist-view-human-origins.aspx (accessed December 1, 2014).

20. Harris Poll, "What People Do and Do Not Believe In."

21. Benjamin Radford, interview with the author, February 9, 2015.

22. Harris Poll, "What People Do and Do Not Believe In."

23. Lyons, "Paranormal Beliefs Come (Super) Naturally to Some."

24. Jayant V. Narlikar, "An Indian Test of Indian Astrology," *Skeptical Inquirer* 37, no. 2 (March/April 2013), http://www.csicop.org/si/show/an_indian _test_of_indian_astrology/ (accessed January 15, 2015).

25. Kelly Beatty, "A Sign of the Times," *Sky and Telescope*, January 20, 2011, http://www.skyandtelescope.com/astronomy-news/observing-news/a-sign -of-the-times/ (accessed January 7, 2015).

26. For an overview of key problems with astrology, see Guy P. Harrison, *50 Popular Beliefs That People Think Are True* (Amherst, NY: Prometheus Books, 2011), pp. 139–43.

27. Joan Quigley, *What Does Joan Say? My Seven Years as White House Astrologer to Nancy and Ronald Reagan* (New York: Birch Lane, 1990).

28. "About St John's College, Oxford—Admissions," https://www.facebook .com/stjohnsoxford/info?tab=overview (accessed June 16, 2015).

29. "Astrology-Loving MP Seeks Health Answers in the Stars," BBC, July 25, 2014, http://www.bbc.com/news/uk-politics-28464009 (accessed January 15, 2015).

30. Dean Buonomano, *Brain Bugs: How the Brain's Flaws Shape Our Lives* (New York: W. W. Norton, 2011), pp. 221–22.

31. Robert T. Carroll, "Edgar Mitchell's ESP Experiment," *Skeptic's Dictionary*, December 27, 2010, http://skepdic.com/edgarmitchell.html (accessed January 12, 2015).

32. Robert Todd Carroll, "Quantum Hologram," *Skeptic's Dictionary*, http://skepdic.com/quantumhologram.html (accessed May 21, 2015).

33. NASA, "Biographical Data: Edgar Dean Mitchell (Captain, USN, Ret.)," http://www.jsc.nasa.gov/Bios/htmlbios/mitchell-ed.html (accessed January 19, 2015).

34. Harris Poll, "What People Do and Do Not Believe In."

35. James Hookway, "Spirits Move Indonesians to Give Shamans the Business," *Wall Street Journal*, November 22, 2012, http://online.wsj.com/ articles/SB10000872396390444592704578066471037376696 (accessed December 8, 2014).

36. Hank Davis, *Caveman Logic: The Persistence of Primitive Thinking in a Modern World* (Amherst, NY: Prometheus Books, 2009), pp. 30, 37.

37. Edward O. Wilson, *The Social Conquest of Earth* (New York: W. W. Norton, 2012), p. 7.

38. Ian Tattersall, *Becoming Human: Evolution and Human Uniqueness* (New York: Harcourt Brace, 1998), pp. 233–34.

39. Ibid.

40. Lauren Cox, "Why Do We Spend $34 Billion in Alternative Medicine?" ABC News, January 29, 2009, http://abcnews.go.com/Health/WellnessNews/story?id=8215703 (accessed November 7, 2014).

41. Jackson Galaxy, "Jackson Galaxy Spirit Essences Bully Remedy Item #55484," https://store.jacksongalaxy.com/store/jg/item/55484/jackson-galaxy-spirit-essences-bully-remedy?source=10-1375590785-1 (accessed June 16, 2015).

42. "Power Balance Come Clean and Offer Refunds after Admitting Wristbands Do Not Enhance Performance," *Telegraph*, January 4, 2011, http://www.telegraph.co.uk/sport/8238255/Power-Balance-come-clean-and-offer-refunds-after-admitting-wristbands-do-not-enhance-performance.html (accessed May 22, 2015).

43. John Saward, "Hallowed Be Thy Name Brand: The Religious Consumerism of Megachurch Pastor Joel Osteen," *Vice*, January 30, 2015, http://www.vice.com/read/hallowed-be-thy-namebrand-the-religious-consumerism-of-megachurch-pastor-joel-osteen (accessed May 12, 2015).

44. Sam Stringer, "Minister Creflo Dollar Asks for $60 Million in Donations for a New Jet," CNN, March 16, 2015, http://www.cnn.com/2015/03/13/living/creflo-dollar-jet-feat/index.html (accessed May 14, 2015).

45. Ryan T. Cragun, Stephanie Yeager, and Desmond Vega, "Research Report: How Secular Humanists (and Everyone Else) Subsidize Religion in the United States," *Free Inquiry* (June/July 2012): 39–46.

46. Chris Hedges, *Empire of Illusion: The End of Literacy and the Triumph of Spectacle* (New York: Nation Books, 2010), p. 44.

47. Ibid.

48. David Aaronovitch, *Voodoo Histories: The Role of the Conspiracy Theory in Shaping Modern History* (New York: Penguin Group, 2010), p. 3.

49. Christine D. Johnson, "'Heaven Is for Real' Hits Major Sales Milestone," *Christian Retailing*, December 11, 2014, http://www.christianretailing.com/index.php/newsletter/latest/27680-heaven-is-for-real-hits-major-sales-milestone (accessed May 23, 2015).

50. "'The Harbinger' Marks a Million in Sales with Two Evangelical Christian Publishers Association Awards," *Christian Retailing*, May 29, 2013, http://www.christianretailing.com/index.php/news/industry-news/25756-the-harbinger-marks-a-million-in-sales-with-two-evangelical-christian-publishers-association-awards (accessed May 23, 2015).

51. Jeremy Burns, "Piper's '90 Minutes in Heaven' Back on 'New York Times' Best-Seller List," *Christian Retailing*, April 25, 2014, http://www.christianretailing.com/index.php/newsletter/latest/26973-pipers-90-minutes-in-heaven-back-on-new-york-times-best-seller-list (accessed May 23, 2015).

52. Sarah Eekhoff Zylstra, "The 'Boy Who Came Back from Heaven' Retracts Story," *Christianity Today*, January 15, 2015, http://www.christianity today.com/gleanings/2015/january/boy-who-came-back-from-heaven-retraction .html?paging=off (accessed February 23, 2015).

53. Guy P. Harrison, *50 Popular Beliefs That People Think Are True* (Amherst, NY: Prometheus Books, 2011), pp. 100–107.

54. Julien Musolino, interview with the author, February 23, 2015.

55. Leonid Rozenblit and Frank Keil, "The Misunderstood Limits of Folk Science: An Illusion of Explanatory Depth," *Cognitive Science* 26, no. 5 (September–October 2002): 521–62.

CHAPTER 2: WHERE DO BRAINS COME FROM?

1. Bob Brier, *Egyptian Mummies: Unraveling the Secrets of an Ancient Art* (New York: Harper Perennial, 1996), pp. 61–62.

2. Michael S. Sweeney, *Brain: The Complete Mind: How It Develops, How It Works, and How to Keep It Sharp* (Washington, DC: National Geographic, 2009), p. 3.

3. Ananya Mandal, "Semen and Culture," *News Medical*, May 27, 2015, http://www.news-medical.net/health/Semen-and-Culture.aspx (accessed May 26, 2015).

4. Denis Noble, Dario DiFrancesco, and Diego Zancani, "Leonardo da Vinci and the Origin of Semen," Royal Society, *Notes and Records*, August 20, 2014, http://rsnr.royalsocietypublishing.org/content/early/2014/08/14/rsnr .2014.0021#xref-fn-7-1 (accessed May 27, 2015).

5. Ibid.

6. Gengo Tanaka, Xianguang Hou, Xiaoya Ma, Gregory D. Edgecombe, and Nicholas J. Strausfeld, "Chelicerate Neural Ground Pattern in a Cambrian Great Appendage Arthropod," *Nature* 502, no. 7471 (2013): 364, doi:10.1038/ nature12520 (accessed November 22, 2015).

7. Seth Shostak, interview with the author, March 28, 2015.

8. Gary Marcus, *Kluge: The Haphazard Construction of the Human Mind* (New York: Houghton Mifflin, 2008), p. 2.

9. Ibid., p. 6.

10. Stephen Jay Gould, *Wonderful Life: The Burgess Shale and the Nature of History* (New York: W. W. Norton, 2007), p. 291.

11. Charles Darwin, *The Descent of Man* (New York: Penguin Classics), p. 18.

12. John Hawks, "How Has the Human Brain Evolved?" *Scientific American*, June 6, 2013, http://www.scientificamerican.com/article/how-has-human -brain-evolved (accessed November 1, 2015).

13. Steven Rose, *The Future of the Brain* (New York: Oxford University Press, 2005), p. 91.

14. Cameron M. Smith, interview with the author, February 11, 2015.

15. Hawks, "How Has the Human Brain Evolved?"

16. J. B. Smaers and C. Soligo, "Brain Reorganization, Not Relative Brain Size, Primarily Characterizes Anthropoid Brain Evolution," *Royal Society Proceedings B*, March 27, 2013, doi:10.1098/rspb.2013.0269, http://rspb.royalsociety publishing.org/content/280/1759/20130269.full (accessed January 19, 2015).

17. Ian Tattersall, *Becoming Human: Evolution and Human Uniqueness* (New York: Harcourt Brace, 1998), pp. 72–73.

18. Steve Bradt, "Invention of Cooking Drove Evolution of the Human Species, New Book Argues," *Harvard Gazette*, June 1, 2009, http://news .harvard.edu/gazette/story/2009/06/invention-of-cooking-drove-evolution-of -the-human-species-new-book-argues/ (accessed February 2, 2015).

19. Joanne Bradbury, "Docosahexaenoic Acid (DHA): An Ancient Nutrient for the Modern Human Brain," *Nutrients*, May 2011, http://www.ncbi.nlm.nih .gov/pmc/articles/PMC3257695/ (accessed January 1, 2015).

20. David Robson, "Sharp Thinking: How Shaping Tools Built Our Brains," *New Scientist*, March 3, 2014, http://www.newscientist.com/article/ mg22129580.600-sharp-thinking-how-shaping-tools-built-our-brains.html# .VJHkmivF_X6 (accessed December 30, 2014).

21. Frederick L Coolidge and Thomas Wynn, "The Effects of the Tree-to-Ground Sleep Transition in the Evolution of Cognition in Early *Homo*," *Before Farming*, April 2006, http://www.uccs.edu/Documents/fcoolidg/Before%20 Farming%202006%20Dream%20paper.pdf (accessed March 26, 2015). See also David Robinson, "The Story in the Stones," *New Scientist*, March 1, 2014, p. 36.

22. Ibid.

23. The sleeping-dolphins-shall-rule-the-world idea came up in a conversation with Kelly Frede.

24. Jim Holt, "The Man behind the Meme," *Slate*, December 1, 2004, http:// www.slate.com/articles/arts/egghead/2004/12/the_man_behind_the_meme .html (accessed January 30, 2015).

25. Rob DeSalle and Ian Tattersall, *The Brain: Big Bangs, Behaviors, and Beliefs* (New Haven and London: Yale University Press, 2012), pp. 26–27.

26. Sid Perkins, "Oldest Primate Skeleton Unveiled," *Nature*, June 5, 2013, http://www.nature.com/news/oldest-primate-skeleton-unveiled-1.13142 (accessed December 5, 2014).

27. Kathleen McAuliffe, "If Modern Humans Are So Smart, Why Are Our Brains Shrinking?" *Discover*, January 20, 2011, http://discover magazine.com/2010/sep/25-modern-humans-smart-why-brain-shrinking (accessed December 3, 2014).

28. Spencer Wells, *Pandora's Seed: The Unforeseen Cost of Civilization* (New York: Random House, 2010).

29. Dean Buonomano, *Brain Bugs: How the Brain's Flaws Shape Our Lives* (New York: W. W. Norton, 2011), p. 16.

CHAPTER 3: EXPLORING YOUR BRAIN

1. Majid Fotuhi, David Do, and Clifford Jack, "Modifiable Factors That Alter the Size of the Hippocampus with Ageing," *Nature Reviews: Neurology*, advance online publication, March 13 2012, pp. 1–2, doi:10.1038/nrneurol.2012.27, http://www.neurexpand.com/wp-content/uploads/2013/08/Modifiable-factors.pdf (accessed January 5, 2015).

2. Ibid., p. 2.

3. J. Brabec, A. Rulseh, B. Hoyt, M. Vizek, D. Horinek, J. Hort, and P. Petrovicky, "Volumetry of the Human Amygdala—An Anatomical Study," *Psychiatry Res.* 182, no. 1 (April 30, 2010): 67–72, doi:10.1016/j.pscychresns.2009.11.005, epub March 15, 2010, http://www.ncbi.nlm.nih.gov/pubmed/20227858 (accessed January 2, 2015).

4. "Amygdala Volume and Social Network Size in Humans," *Nature Neuroscience* 14 (2011; published online December 26, 2010): 163–64, doi:10.1038/nn.2724, http://www.nature.com/neuro/journal/v14/n2/abs/nn.2724.html (accessed December 19, 2014).

5. Louis Bergeron, "Size, Connectivity of Brain Region Linked to Anxiety Level in Young Children, Study Shows," Stanford Medicine News Center, November 20, 2014, http://med.stanford.edu/news/all-news/2013/11/size-connectivity-of-brain-region-linked-to-anxiety-level-in-young-children-study-shows.html (accessed January 22, 2015).

6. D. A. Pardini, A. Raine, K. Erickson, and R. Loeber, "Lower Amygdala Volume in Men Is Associated with Childhood Aggression, Early Psychopathic Traits, and Future Violence," *Biological Psychiatry* 75, no. 1 (January 2014): 73–80.

7. A. Abigail Marsh, Sarah A. Stoycos, Kristin M. Brethel-Haurwitz, Paul Robinson, John W. VanMeter, and Elise M. Cardinale, "Neural and Cognitive Characteristics of Extraordinary Altruists," *PNAS* 111, no. 42 (October 21, 2014): 15036–41.

8. S. Binks, D. Chan, and N. Medford, "Abolition of Lifelong Specific Phobia: A Novel Therapeutic Consequence of Left Mesial Temporal Lobectomy," *Neurocase*, January 27, 2014, pp. 79–84.

9. Rachel Feltman, "Meet the Woman Who Can't Feel Fear," *Washington Post*, January 20, 2015, http://www.washingtonpost.com/news/speaking-of-science/wp/2015/01/20/meet-the-woman-who-cant-feel-fear/ (accessed January 22, 2015).

10. Donna Hart, *Man the Hunted: Primates, Predators, and Human Evolution* (New York: Basic Books, 2005).

11. Floyd E. Bloom, ed., *Best of the Brain from Scientific American: Mind, Matter, and Tomorrow's Brain* (New York: Dana, 2007), p. 92.

12. Ibid., pp. 92–93.

13. Robert A. Barton and Chris Vinditti, "Rapid Evolution of the Cerebellum in Humans and Other Great Apes," *Current Biology* 24, no. 20 (October 20, 2014), pp. 2440–44.

14. Jim Bower, interview with the author, January 28, 2015.

15. Bloom, *Best of the Brain from Scientific American*, p. 95.

16. Daniel Levitin, *This Is Your Brain on Music: The Science of a Human Obsession* (New York: Penguin, 2006), pp. 174–75.

17. Bloom, *Best of the Brain from Scientific American*, p. 93.

18. Ibid., pp. 100–101.

19. Ibid.

20. Michael O'Shea, "The Human Brain," *New Scientist* 218, no. 2911(April 6–12, 2013): iv.

21. Brenda Patoine, "The Prefrontal Cortex and Frontal Lobe Disorders: An Interview with Jordan Grafman, PhD," Dana Foundation, January, 2006, http://www.dana.org/Publications/ReportDetails.aspx?id=44153 (accessed June 21, 2015).

22. Ibid.

23. F. A. Azevedo, L. R. Carvalho, L. T. Grinberg, J. M. Farfel, R. E. Ferretti, R. E. Leite, W. Jacob Filho, R. Lent, and S. Herculano-Houzel, "Equal Numbers of Neuronal and Nonneuronal Cells Make the Human Brain an Isometrically Scaled-Up Primate Brain," *Journal of Comparative Neurology* 513, no. 5 (April 10, 2009): 532–41, http://www.ncbi.nlm.nih.gov/pubmed?Db=pubmed&Cmd=ShowDetailView&TermToSearch=19226510 (accessed January 19, 2015).

24. Paul E. Bendheim, *The Brain Training Revolution: A Proven Workout for Healthy Brain Aging* (Naperville, IL: Sourcebooks, 2009), p. 9.

25. Ferris Jabr, "Know Your Neurons: What Is the Ratio of Glia to Neurons in the Brain?" *Brain Waves* (*Scientific American* blog), June 13, 2012, http://blogs.scientificamerican.com/brainwaves/know-your-neurons-what-is-the-ratio-of-glia-to-neurons-in-the-brain/ (accessed May 14, 2015).

26. R. Douglas Fields, "Neuroscience: Map the Other Brain," *Nature* (September 4, 2013), http://www.nature.com/news/neuroscience-map-the-other-brain-1.13654 (accessed December 1, 2014).

27. Ibid.

28. Sebastian Seung, *Connectome: How the Brain's Wiring Makes Us Who We Are* (New York: Houghton Mifflin Harcourt, 2012), p. ix.

29. V. S. Ramachandran, *The Tell-Tale Brain: A Neuroscientist's Quest for What Makes Us Human* (New York: W. W. Norton, 2011), p. 14.

30. John J. Ratey, *Spark: The Revolutionary New Science of Exercise and the Brain* (New York: Little, Brown, 2013), pp. 35–36.

31. Peter S. Eriksson, Ekaterina Perfilieva, Thomas Björk-Eriksson, Ann-Marie Alborn, Claes Nordborg, Daniel A. Peterson, and Fred H. Gage, "Neurogenesis in the Adult Human Hippocampus," *Nature Medicine* 4, no. 11 (November 1998).

32. All quotes in the numbered list below are from Bendheim, *Brain Training Revolution*, pp. 142–44.

CHAPTER 4: WHO'S MINDING THE BRAIN?

1. Barry Bogin, *Patterns of Human Growth* (Cambridge, UK: Cambridge University Press, 1999).

2. Charles Platt, "Future Tech: A Unique Way to Save Your Brain during a Heart Attack," *Discover*, October 1, 2001, http://discovermagazine.com/2001/oct/feattech (accessed June 17, 2015).

3. Quoted in Stuart Wolpert, "Scientists Learn How What You Eat Affects Your Brain—and Those of Your Kids," UCLA Newsroom, July 9, 2008, http://newsroom.ucla.edu/releases/scientists-learn-how-food-affects-52668 (accessed January 19, 2015).

4. Ibid.

5. Ibid.

6. Frank Sacks, "Ask the Expert: Omega-3 Fatty Acids," Nutrition Source, http://www.hsph.harvard.edu/nutritionsource/omega-3/ (accessed January 1, 2015).

7. Federation of American Societies for Experimental Biology (FASEB), "Eating Green Leafy Vegetables Keeps Mental Abilities Sharp," *ScienceDaily*, March 30, 2015, www.sciencedaily.com/releases/2015/03/150330112227.htm (accessed April 5, 2015).

8. Paul E. Bendheim, *The Brain Training Revolution: A Proven Workout for Healthy Brain Aging* (Naperville, IL: Sourcebooks, 2009), p. 59.

9. Lucia Kerti, A. Veronica Witte, Angela Winkler, Ulrike Grittner, Dan Rujescu, and Agnes Flöel, "Higher Glucose Levels Associated with Lower Memory and Reduced Hippocampal Microstructure," *Neurology* 81, no. 20 (November 12, 2013): 1746–752, http://www.neurology.org/content/81/20/1746 (accessed January 2, 2015).

10. Nicole M. Avena, Pedro Rada, and Bartley G. Hoebel, "Evidence for Sugar Addiction: Behavioral and Neurochemical Effects of Intermittent, Excessive Sugar Intake," *Neuroscience & Biobehavioral Reviews* 32, no. 1 (2008): 20–39, http://www.ncbi.nlm.nih.gov/pmc/articles/PMC2235907/ (accessed December 5, 2014).

11. K. A. Page, O. Chan, J. Arora, R. Belfort-Deaguiar, J. Dzuira, B. Roehmholdt, G. W. Cline, S. Naik, R. Sinha, R. T. Constable, and R. S.

Sherwin, "Effects of Fructose vs. Glucose on Regional Cerebral Blood Flow in Brain Regions Involved with Appetite and Reward Pathways," *JAMA* 309, no. 1 (January 2, 2013): 63–70, doi:10.1001/jama.2012.116975, http://www.ncbi .nlm.nih.gov/pubmed/23280226 (accessed January 19, 2015); "What Makes Fructose Fattening? Some Answers Found in the Brain," *Science Daily*, February 9, 2011, http://www.sciencedaily.com/releases/2011/02/110209131951 .htm (accessed December 31, 2015).

12. Helen Briggs, "WHO: Daily Sugar Intake 'Should Be Halved,'" BBC, March 5, 2014, http://www.bbc.com/news/health-26449497 (accessed November 11, 2014).

13. "Sugar 101," American Heart Association, http://www.heart.org/ HEARTORG/GettingHealthy/NutritionCenter/HealthyEating/Sugar-101 _UCM_306024_Article.jsp# (accessed December 22, 2014).

14. Economic Research Services, US Department of Agriculture, "Dietary Assessment of Major Trends in US Food Consumption, 1970–2005," March 2008, http://www.ers.usda.gov/media/210677/eib33_reportsummary_1_.pdf (accessed January 19, 2015).

15. "Drinking Sugar-Sweetened Beverages during adolescence Impairs Memory, Animal Study Suggests," *Science Daily*, July 29, 2014, http://www .sciencedaily.com/releases/2014/07/140729224906.htm (accessed January 2, 2015).

16. Leslie Ridgeway, "Soda Consumers May Be Drinking More Harmful Sugar Than Labels Reveal," *USC News*, June 24, 2014, http://news.usc .edu/63696/soda-consumers-may-be-drinking-more-harmful-sugar-than-labels -reveal/ (accessed January 19, 2015).

17. K. S. Krabbe, A. R. Nielsen, R. Krogh-Madsen, P. Plomgaard, P. Rasmussen, C. Erikstrup, C. P. Fischer, B. Lindegaard, A. M. Petersen, S. Taudorf, N. H. Secher, H. Pilegaard, H. Bruunsgaard, and B. K. Pedersen, "Brain-Derived Neurotrophic Factor (BDNF) and Type 2 Diabetes," *Diabetologia* 50, no. 2 (February 2007): 431–38.

18. R. Molteni, R. J. Barnard, Z. Ying, C. K. Roberts, and F. Gómez-Pinilla, "A High-Fat, Refined Sugar Diet Reduces Hippocampal Brain-Derived Neurotrophic Factor, Neuronal Plasticity, and Learning," *Neuroscience* 112, no. 4 (2002):803–14, http://www.ncbi.nlm.nih.gov/pubmed/12088740 (accessed December 4, 2015).

19. "Lack of Exercise Responsible for Twice as Many Deaths as Obesity," University of Cambridge Research, January 14, 2015, http://ajcn.nutrition.org/content/ early/2015/01/14/ajcn.114.100065.full.pdf+html (accessed January 15, 2015).

20. H. Arem, S. C. Moore, A. Patel, P. Hartge, A. Berrington de Gonzalez, K. Visvanathan, P. T. Campbell, M. Freedman, E. Weiderpass, H. O. Adami, M. S. Linet, I. M. Lee, and C. E. Matthews, "Leisure Time Physical Activity and Mortality: A Detailed Pooled Analysis of the Dose-Response Relationship," *JAMA Internal Medicine* 175, no. 6 (June 1, 2015): 959–67, doi: 10.1001/ jamainternmed.2015.0533 (accessed June 19, 2015).

21. Office of Disease Prevention and Health Promotion, "2008 Physical Activity Guidelines for Americans," http://www.health.gov/paguidelines/guidelines/ (accessed June 19, 2015).

22. Arem et al., "Leisure Time Physical Activity and Mortality."

23. Ibid.

24. ASU News, "Study Shows Resistance Training Benefits Cardiovascular Health," November 29, 2010, http://www.news.appstate.edu/2010/11/29/study-shows-resistance-training-benefits-cardiovascular-health/ (accessed June 17, 2015); Véronique A. Cornelissen, Robert H. Fagard, Ellen Coeckelberghs, and Luc Vanhees, "Impact of Resistance Training on Blood Pressure and Other Cardiovascular Risk Factors: A Meta-Analysis of Randomized, Controlled Trials," *Hypertension* 58 (November 2011): 950–58, doi: 10.1161/HYPERTENSIONAHA.111.177071.

25. University of British Columbia Public Affairs, "How Exercise Can Boost Brain Power," February 21, 2014, http://www.centreforbrainhealth.ca/news/2014/02/21/how-exercise-can-boost-brain-power (accessed February 1, 2015).

26. John J. Ratey, *Spark: The Revolutionary New Science of Exercise and the Brain* (New York: Little, Brown, 2013), p. 242.

27. Aviroop Biswas, Paul I. Oh, Guy E. Faulkner, Ravi R. Bajaj, Michael A. Silver, Marc S. Mitchell, and David A. Alter, "Sedentary Time and Its Association with Risk for Disease Incidence, Mortality, and Hospitalization in Adults: A Systematic Review and Meta-analysis," *Annals of Internal Medicine* (2015), doi:10.7326/M14-1651 (accessed February 9, 2015); E. G. Wilmot, C. L. Edwardson, F. A. Achana, M. J. Davies, T. Gorely, L. J. Gray, K. Khunti, T. Yates, and S. J. H. Biddle, "Sedentary Time in Adults and the Association with Diabetes, Cardiovascular Disease and Death: Systematic Review and Meta-analysis," *Diabetologia* 55, no. 11 (2012): 2895, doi:10.1007/s00125-012-2677-z (accessed February 14, 2015).

28. *Science Daily*, "Sitting for Long Periods Increases Risk of Disease and Early Death, Regardless of Exercise," January 19, 2015, http://www.sciencedaily.com/releases/2015/01/150119171701.htm (accessed February 14, 2015).

29. Matthew Ricard, Antoine Lutz, and Richard J. Davidson, "Mind of the Meditator," *Scientific American*, November 2014, pp. 38–45.

30. Bendheim, *Brain Training Revolution*, p. 53.

31. Brenda Patoine, "Move Your Feet, Grow New Neurons?" Dana Foundation, May 2007, http://www.dana.org/Publications/Brainwork/Details.aspx?id=43678 (accessed December 20, 2015).

32. Ibid.

33. David Eagleman, *Incognito: The Secret Lives of the Brain* (New York: Vintage, 2012), pp. 128–29.

34. World Health Organization, "Depression," Fact Sheet No. 369, October 2012, http://www.who.int/mediacentre/factsheets/fs369/en/ (accessed December 12, 2012).

35. Louise B. Andrew, "Depression and Suicide," *Medscape*, August 5, 2014, http://emedicine.medscape.com/article/805459-overview (accessed January 15, 2015).

36. Ibid.

37. Oliver Renick, "France, US Have Highest Rates of Depression Rates in World, Study Suggests," *Bloomberg News*, July 25, 2011, http://www.bloomberg .com/news/2011-07-26/france-u-s-have-highest-depression-rates-in-world-study -suggests.html (accessed November 22, 2015).

38. Andrew, "Depression and Suicide."

39. World Health Organization, "Depression."

40. "Moderate Exercise Not Only Treats, But Prevents Depression," University of Toronto, October 25, 2013, http://media.utoronto.ca/media-releases/ moderate-exercise-not-only-treats-but-prevents-depression/ (accessed January 15, 2015).

41. K. A. Barbour, T. M. Edenfield, and J. A. Blumenthal, "Exercise as a Treatment for Depression and Other Psychiatric Disorders: A Review," *Journal of Psychosomatic Medicine* 27, no. 6 (November/December 2007): 359–67, doi:10.1097/01.HCR.0000300262.69645.95.

42. Ratey, *Spark*, pp. 122–23, 135.

43. Katie Moisse, "5 Health Hazards Linked to Lack of Sleep," June 11, 2012, http://abcnews.go.com/Health/Sleep/health-hazards-linked-lack-sleep/ story?id=16524313 (accessed November 2, 2014).

44. Fisher Center for Alzheimer's Research Foundation, "Poor Sleep May Be Linked to Alzheimer's Disease," http://www.alzinfo.org/articles/poor-sleep -may-be-linked-to-alzheimers-disease/ (accessed January 30, 2015).

45. Bendheim, *Brain Training Revolution*, p. 256.

46. Ibid., p. 256.

47. M. S. Christian and A. P. Ellis, "Examining the Effects of Sleep Deprivation on Workplace Deviance: A Self-Regulatory Perspective," *Academy of Management Journal* 54, no. 5 (2011): 913–34.

48. Bendheim, *Brain Training Revolution*, p. 257.

49. Guang Yang, Cora Sau Wan Lai, Joseph Cichon, Lei Ma, Wei Li, Wen-Biao Gan, "Sleep Promotes Branch-Specific Formation of Dendritic Spines after Learning," *Science* 344, no. 6188 (June 6, 2014): 1173–78, http://www. sciencemag.org/content/344/6188/1173 (accessed January 3, 2015).

50. Bendheim, *Brain Training Revolution*, p. 266.

51. Steven J. Frenda, Lawrence Patihis, Elizabeth F. Loftus, Holly C. Lewis, Kimberly M. Fenn, "Sleep Deprivation and False Memories," *Psychological Science*. 25 no. 9 (September 2014): 1674–81.

52. Jon Hamilton, "Sleep's Link to Learning and Memory Traced to Brain Chemistry," NPR, November 20, 2014, http://www.npr.org/blogs/ health/2014/11/20/365213989/sleeps-link-to-learning-and-memory-traced-to -brain-chemistry (accessed October 13, 2014).

53. Caroline Williams, "A User's Guide to the Mind," *New Scientist: The Collection—The Human Brain*, vol. 2, no. 1, p. 121.

54. Mareike B. Wieth and Rose T. Zacks, "Time of Day Effects on Problem Solving: When the Non-optimal Is Optimal," *Thinking and Reasoning* 17, no. 4 (2011): 387–401.

CHAPTER 5: BRINGING HUMAN VISION INTO FOCUS

1. NASA News, "UFO Planet," NASA, May 3, 2004, http://science.nasa.gov/science-news/science-at-nasa/2004/03may_maximumvenus/ (accessed January 1, 2015).

2. Ronald A. Rensink, J. Kevin O'Regan, and James J. Clark, "To See or Not to See: The Need for Attention to Perceive Changes in Scenes," *Psychological Science* 8, no. 5 (September 1997): 368–73.

3. Yasmin Anwar, "Hit a 95 mph Baseball? Scientists Pinpoint How We See It Coming," UC Berkley News Center, May 8, 2013, http://newscenter.berkeley.edu/2013/05/08/motion-vision/ (accessed December 30, 2014); Gerrit W. Maus, Jason Fischer, and David Whitney, "Motion-Dependent Representation of Space in Area MT+," *Neuron* 78, no. 3 (May 8, 201): 554–62, http://www.sciencedirect.com/science/article/pii/S0896627313002572 (accessed December 30, 2014).

4. I recommend *The Invisible Gorilla: How Our Intuitions Deceive Us*, by Christopher Chabris and Daniel Simons (New York: Crown, 2010). Test your friends' inattentional blindness with the short video at www.theinvisiblegorilla.com/videos.html.

5. MedLinePlus, "Hallucinations," National Institutes of Health, December 24, 2014, http://www.nlm.nih.gov/medlineplus/ency/article/003258.htm (accessed December 7, 2014).

6. V. S. Ramachandran, *The Tell-Tale Brain: A Neuroscientist's Quest for What Makes Us Human* (New York: W. W. Norton, 2011), p. 229.

CHAPTER 6: DON'T FORGET HOW MEMORY WORKS

1. D. J. Simons, C. F. Chabris, "What People Believe about How Memory Works: A Representative Survey of the U.S. Population," *PLoS ONE* 6, no. 8 (August 3, 2011): e22757, doi:10.1371/journal.pone.0022757, http://www.plosone.org/article/info%3Adoi%2F10.1371%2Fjournal.pone.0022757 (accessed January 31, 2015).

2. Gary Marcus, *Kluge: The Haphazard Construction of the Human Mind* (New York: Houghton Mifflin, 2008), p. 22.

3. Ibid., p. 21.

4. Ibid., p. 23.

5. Sandra Aamodt and Sam Wang, *Welcome to Your Brain: Why You Lose Your Car Keys but Never Forget How to Drive, and Other Puzzles of Everyday Life* (New York: Bloomsbury, 2008), p. 82.

6. Association for Psychological Science, "Boosting Memory with Wakeful Resting," July 23, 2012, http://www.psychologicalscience.org/index.php/news/releases/boosting-new-memories-with-wakeful-resting.html (accessed April 30, 2015).

7. Joannie Schrof Fischer, "What Is Memory Made Of?" *Mysteries of Science (U.S. News and World Report)*, 2002, p. 27.

8. Marcus, *Kluge*, p. 39.

9. Dean Buonomano, *Brain Bugs: How the Brain's Flaws Shape Our Lives* (New York: W. W. Norton, 2011), p. 68.

10. Elizabeth Loftus, "Leading Questions and the Eyewitness Report," *Cognitive Psychology*, October 31, 1975, p. 562, https://webfiles.uci.edu/eloftus/CognitivePsychology75.pdf (accessed January 1, 2015).

11. Ibid.

12. Elizabeth Loftus, "The Reality of Repressed Memories," *American Psychologist* 48, no. 5 (May 1993): 524.

13. Mahzarin R. Banaji and Anthony G. Greenwald, *Blind Spot: Hidden Biases of Good People* (New York: Delacorte, 2013), p. 9.

14. Elizabeth Loftus, "Elizabeth Loftus: The Fiction of Memory," 17:36, TED talk filmed June 2013, https://www.ted.com/talks/elizabeth_loftus_the_fiction_of_memory#t-316982 (accessed January 6, 2015).

15. D. Gallo, H. Roediger, and K. McDermott, "Associative False Recognition Occurs without Strategic Criterion Shifts," *Psychonomic Bulletin and Review* 8, no. 3 (2001): 579–86.

16. National Research Council of the National Academies; Committee on Scientific Approaches to Understanding and Maximizing the Validity and Reliability of Eyewitness Identification in Law Enforcement and the Courts; Committee on Science, Technology, and Law; Policy and Global Affairs; Committee on Law and Justice; Division of Behavioral and Social Sciences and Education, *Identifying the Culprit: Assessing Eyewitness Identification* (Washington, DC: National Academies, 2014), p. 22.

17. "DNA Exoneree Case Profiles," Innocence Project, http://www.innocenceproject.org/know/ (accessed January 6, 2015).

18. "Eyewitness Identification," Innocence Project, http://www.innocenceproject.org/fix/Eyewitness-Identification.php (accessed January 6, 2015).

19. National Research Council of the National Academies et al., *Identifying the Culprit*.

20. Ibid.

21. Benjamin Weiser, "In New Jersey, Rules Are Changed on Witness

IDs," *New York Times*, August 24, 2011, http://www.nytimes.com/2011/08/25/
nyregion/in-new-jersey-rules-changed-on-witness-ids.html (accessed January
4, 2015).

22. Benjamin Weiser, "New Jersey Court Issues Guidance for Juries
about Reliability of Eyewitnesses," *New York Times*, July 19, 2012, http://www
.nytimes.com/2012/07/20/nyregion/judges-must-warn-new-jersey-jurors-about
-eyewitnesses-reliability.html?_r=2&ref=nyregion& (accessed January 19, 2015).

23. Jennifer M. Talarico and David C. Rubin, "Confidence, Not Consistency, Characterizes Flashbulb Memories," *Psychological Science* 14, no. 5
(September 2003).

24. T. Sharot, E. A. Martorella, M. R. Delgado, and E. A. Phelps, "How
Personal Experience Modulates the Neural Circuitry of Memories of September 11," *Proceedings of the National Academy of Sciences* 104, no. 1
(USA, 2007):389–94, http://www.ncbi.nlm.nih.gov/pmc/articles/PMC1713166/
(accessed January 16, 2015).

25. J. S. Simons, interview with the author, January 27, 2015.

CHAPTER 7: THE SHADOW BRAIN

1. Benjamin Libet, Curtis Gleason, Eldwood Wright, and Dennis Pearl,
"Time of Conscious Intention to Act in Relation to Onset of Cerebral Activity
(Readiness Potential)," *Brain* 106, no. 3 (September 1983): 623–42.

2. John A. Bargh, "Our Unconscious Mind," *Scientific American*, January
2014, p. 32.

3. David Eagleman, *Incognito: The Secret Lives of the Brain* (New York:
Vintage, 2012), p. 9.

4. Bargh, "Our Unconscious Mind," p. 37.

5. Gillian Rhodes, Fiona Proffitt, Jonathon M. Grady, and Alex Sumich,
"Facial Symmetry and the Perception of Beauty," *Psychonomic Bulletin &
Review* 5, no. 4 (December 1998): 659–69.

6. Eagleman, *Incognito*, pp. 5–6.

7. Erik von Däniken, *Chariots of the Gods?* (New York: Berkley Publishing Group, 1999), copyright page.

8. Jamie Tarabay, "In Death, Saddam Fascinates Iraqi Sup-
porters," NPR, January 8, 2007, http://www.npr.org/templates/story/story
.php?storyId=6749072 (accessed January 22, 2015).

9. "Woman 'Blessed by the Holy Toast,'" BBC, November 17, 2004, http://
news.bbc.co.uk/2/hi/americas/4019295.stm (accessed June 4, 2015).

10. Cara Hutt, "Is That You, Jesus . . . In a Tortilla?" Headline News
(CNN), December 14, 2014, http://www.hlntv.com/article/2014/12/11/jesus
-tortilla (accessed January 23, 2015).

11. Mary Jane Park, "An Image of Jesus: All That in a Bag of Chips," *Tampa Bay Times*, June 8, 2005, http://www.sptimes.com/2005/06/08/Neighborhoodtimes/An_image_of_Jesus__Al.shtml (accessed January 21, 2015).

12. Fox 10 Phoenix, "Image of Jesus Seen on Moth's Wings," September 5, 2014, http://www.fox10phoenix.com/story/26462126/2014/09/05/image-of -jesus-seen-on-moths-wings (accessed February 18, 2015).

13. "Virgin Mary Grilled Cheese vs. Jesus Fish Stick," *Chicago Tribune*, November 26, 2004.

14. "Hordes Flocking to See 'Miracle Lamb,'" BBC, March 26, 2004, http://news.bbc.co.uk/2/hi/middle_east/3572325.stm (accessed December 18, 2015).

15. "Löwen Rufen Name Allah," YouTube video, 0:15, video of lion allegedly saying "Allah," posted by Musliman, March 8, 2008, https://www.youtube.com/watch?v=uWwVpv39ifc (accessed February 18, 2015).

16. Cheryl Eddy, "ET's Face Appears in Tree Stump," *io9*, January 21, 2015, http://io9.com/e-t-s-face-appears-in-tree-stump-1680986163 (accessed February 18, 2015).

17. Stephen Jay Gould, *The Flamingo's Smile: Reflections in Natural History* (New York: W. W. Norton, 1985), p. 199.

18. Hank Davis, *Caveman Logic: The Persistence of Primitive Thinking in a Modern World* (Amherst, NY: Prometheus Books, 2009), pp. 26–27.

19. Michael Shermer, *The Believing Brain: From Ghosts and Gods to Politics and Conspiracies How We Construct Beliefs and Reinforce Them as Truths* (New York: Times Books, 2011), p. 5.

20. Michael Shermer, "Why People Believe Invisible Agents Control the World," *Scientific American*, May 18, 2009, http://www.scientificamerican.com/article/skeptic-agenticity/ (accessed December 19, 2014).

CHAPTER 8: THE ENEMY WITHIN

1. Read the Anthropological Association's statement on race here: "American Anthropological Association Statement on 'Race,'" May 17, 1998, www.aaanet.org/stmts/racepp.htm (accessed May 21, 2015).

2. Jennifer L. Doleac and Luke C. D. Stein, "The Visible Hand: Race and Online Market Outcomes," *Economic Journal* 123, no. 572 (November 2013): F469–F492.

3. Joshua Correll, Bernadette Park, Charles M. Judd, Bernd Wittenbrink, Melody S. Sadler, and Tracie Keesee, "Across the Thin Blue Line: Police Officers and Racial Bias in the Decision to Shoot," *Journal of Personality and Social Psychology* 92, no. 6 (June 2007): 1006–23.

4. John F. Dovidio, Samuel L. Gaertner, and Mark P. Zanna, eds., "Aver-

sive Racism," 2004, *Advances in Experimental Social Psychology* 36 (2004): 1–52.

   5. Steven J. Spencer, Claude M. Steele, and Diane M. Quinn, "Stereotype Threat and Women's Math Performance," *Journal of Experimental Social Psychology* 35, no. 1 (1999): 4–28.

   6. Claude M. Steele and Joshua Aronson, "Stereotype Threat and the Intellectual Test Performance of African Americans," *Journal of Personality and Social Psychology* 69, no. 5 (November 1995): 797–811.

   7. Ibid.

   8. Jeff Stone, "Battling Doubt by Avoiding Practice: The Effects of Stereotype Threat on Self-Handicapping in White Athletes," *Personality and Social Psychology Bulletin* 28, no. 12 (December 2002): 1667–78.

   9. For an excellent overview of the IAT, see Mahzarin R. Banaji and Anthony G. Greenwald, *Blind Spot: Hidden Biases of Good People* (New York: Delacorte, 2013). To take the test yourself, go to the Project Implicit website at https://implicit.harvard.edu/implicit/index.jsp.

   10. Elizabeth A. Phelps, Jennifer T. Kubota, Jian Li, Eyal Bar-David, and Mahzarin R. Banaji, "The Price of Racial Bias: Intergroup Negotiations in the Ultimatum Game," *Psychological Science* 24, no. 12 (December 2013): 2498–2504.

   11. Janice A. Sabin, Maddalena Marini, and Brian A. Nosek, "Implicit and Explicit Anti-Fat Bias among a Large Sample of Medical Doctors by BMI, Race/Ethnicity and Gender," *PLoS ONE* 7, no. 11 (2012): e48448, doi:10.1371/journal.pone.0048448 (accessed February 3, 2015).

   12. Elizabeth N. Chapman, Anna Kaatz, and Molly Carnes, "Physicians and Implicit Bias: How Doctors May Unwittingly Perpetuate Health Care Disparities," *Journal of General Internal Medicine* 28, no. 11 (November 2013): 1504–10.

   13. Project Implicit, "Frequently Asked Questions," https://implicit.harvard.edu/implicit/faqs.html (accessed January 25, 2015).

   14. Christopher L. Aberson, Carl Shoemaker, and Christina Tomolillo, "Implicit Bias and Contact: The Role of Interethnic Friendships," *Journal of Social Psychology* 144, no. 3 (2004): 335–47.

   15. Project Implicit, https://implicit.harvard.edu/implicit/index.jsp (accessed June 15, 2015).

   16. Mahzarin Banaji, "Unraveling Beliefs," in *What Are You Optimistic About?* ed. Jon Brockman, pp. 266–68 (New York: Harper Collins, 2007).

   17. Irene V. Blair, Jennifer E. Ma, and Alison P. Lenton, "Imagining Stereotypes Away: The Moderation of Implicit Stereotypes through Mental Imagery," *Journal of Personality and Social Psychology* 81, no. 5 (2001): 828–41.

   18. Mahzarin R. Banaji and Anthony G. Greenwald, *Blind Spot: Hidden Biases of Good People* (New York: Delacorte, 2013), p. 167.

   19. Mahzarin Banaji, interview with author, February 14, 2015.

CHAPTER 9: THREE CRAZY THINGS THAT LIVE IN YOUR HEAD

1. Francis Bacon quoted in Raymond Nickerson, "Confirmation Bias: A Ubiquitous Phenomenon in Many Guises," *Review of General Psychology* 2, no. 2 (1998): 176.

2. Hank Davis, *Caveman Logic: The Persistence of Primitive Thinking in a Modern World* (Amherst, NY: Prometheus Books, 2009), pp. 183–84.

3. Chanel Prabatah, interview with the author, March 22, 2015.

4. Guy P. Harrison, "God Is in This Place," *Caymanian Compass*, November 19, 1993, pp. 10–11.

5. Daniel Sokal, "Inside the Mind of the Doctor," BBC, May 9, 2007, http://www.news.bbc.co.uk/2/hi/health/6610719.stm (accessed July 12, 2015).

6. A. Tversky and D. Kahneman, "Judgment under Uncertainty: Heuristics and Biases," *Science* 185, no. 4157(1974): 1124–31, doi:10.1126/science.185.4157.1124.

7. Ibid.

8. Robert Sapolsky, *The Mind: Leading Scientists Explore the Brain, Memory, Personality, and Happiness*, ed. John Brockman (New York: Harper Perennial, 2011), pp. 191–99; Colin Barras, "The Cat Made Me Do It: Is Your Pet Messing with Your Mind?" *New Scientist*, May 29, 2015, http://www.newscientist.com/article/mg22630230.200-the-cat-made-me-do-it-is-your-pet-messing-with-your-mind.html#.VYeRBkJRFjo (accessed June 21, 2015); Kathleen Maucaliffe, "How Your Cat Is Making You Crazy," *Atlantic*, http://www.theatlantic.com/magazine/archive/2012/03/how-your-cat-is-making-you-crazy/308873/ (accessed June 4, 2015).

9. Barras, "Cat Made Me Do It."

10. Alanna Collen, *10 % Human: How Your Body's Microbes Hold the Key to Health and Happiness* (New York: Harper, 2015), p. 96.

11. Barras, "Cat Made Me Do It."

12. American Association for Psychological Science, "I Knew It All Along . . . Didn't I?—Understanding Hindsight Bias," September 6, 2012, http://www.psychologicalscience.org/index.php/news/releases/i-knew-it-all-along-didnt-i-understanding-hindsight-bias.html (accessed February 22, 2012).

13. Ibid.

CHAPTER 10: AN ALTERNATE VIEW OF ALTERNATIVE MEDICINE

1. Paul A. Offit, *Do You Believe in Magic? Vitamins, Supplements, and All Things Natural: A Look behind the Curtain* (New York: Harper Paperbacks, 2014), p. 42.

2. Paul A. Offit, "Alternative Healing or Quackery?" CNN, June 18, 2013,

http://www.cnn.com/2013/06/18/health/alternative-medicine-offit/ (accessed February 5, 2015).

3. Michael Bond, "Interview: The Complementary Medicine Detective," *New Scientist*, April 26, 2008, http://www.newscientist.com/article/mg19826531.400-interview-the-complementary-medicine-detective.html?full =true&print=true#.VOE8EfnF_X4 (accessed February 4, 2015).

4. Ibid.

5. Paul Offit, "Alternative Healing or Quackery?" CNN, June 18, 2013, http://www.cnn.com/2013/06/18/health/alternative-medicine-offit/ (accessed February 5, 2015).

6. Consumer Reports, "Alternative Treatments," July 2011, http://www.consumerreports.org/cro/2012/04/alternative-treatments/index.htm (accessed January 4, 2015).

7. David Cohen, "Alternative Medicine Investigator: Placebos and Platitudes," August 23, 2011, http://www.newscientist.com/article/mg21128260.300-alternative-medicine-investigator-placebos-and-platitudes.html#.VOE1l_nF_X4 (accessed February 15, 2015).

8. Joseph Stromberg and Sarah Zielinski, "Ten Threatened and Endangered Species Used in Traditional Medicine," Smithsonian.com, October 18, 2011, http://www.smithsonianmag.com/science-nature/ten-threatened-and-endangered-species-used-in-traditional-medicine-112814487/ (accessed June 7, 2015).

9. David Cohen, "Alternative Medicine Investigator: Placebos and Platitudes," *New Scientist*, August 23, 2011, http://www.newscientist.com/article/mg21128260.300-alternative-medicine-investigator-placebos-and-platitudes.html#.VOE1l_nF_X4 (accessed February 15, 2015).

10. Rose Shapiro, *Suckers: How Alternative Medicine Makes Fools of Us All* (London: Harvill Secker, 2008), p. 1.

11. Ibid.

12. Raine Sihvonen, Mika Paavola, Antti Malmivaara, Ari Itälä, Antti Joukainen, Heikki Nurmi, Juha Kalske, and Teppo L. N. Järvinen, "Arthroscopic Partial Meniscectomy versus Sham Surgery for a Degenerative Meniscal Tear," *New England Journal of Medicine* 369 (December 26, 2013): 2515–24, doi: 10.1056/NEJMoa1305189; Aaron E. Carroll, "The Placebo Effect Doesn't Apply Just to Pills," *New York Times*, October 6, 2014, http://www.nytimes.com/2014/10/07/upshot/the-placebo-effect-doesnt-apply-just-to-pills.html?_r=0&abt=0002&abg=0 (accessed June 7, 2015).

13. Jonathan C. Smith, *Pseudoscience and Extraordinary Claims of the Paranormal: A Critical Thinker's Toolkit* (West Sussex, UK: Wiley-Blackwell, 2010), p. 186.

14. Ibid.

15. John Byrne, e-mail interview with the author, February 18, 2013.

16. Osher Center for Integrative Medicine, "Networking Modern Science

with Integrative Approaches to Wellness," http://brighamandwomens.org/
Departments_and_Services/medicine/services/oshercenter/default.aspx (accessed June 16, 2015); Kimball Atwood, "Harvard Medical School: Veritas for Sale (Part I)," *Science-Based Medicine*, https://www.sciencebasedmedicine.org/harvard-medical-school-veritas-for-sale-part-i/ (accessed June 16, 2015).

17. Marilyn Marchione, "Many Herbal Supplements Made by Big Pharma," NBC News, June 9, 2009, http://www.nbcnews.com/id/31188920/ns/health-alternative_medicine/t/many-herbal-products-made-big-pharma/#.VYMim0JRHIU (accessed June 18, 2015).

18. Jackie Northam, "New York State Clamps Down on Herbal Supplements," NPR, February 3, 2015, http://www.npr.org/blogs/thetwo-way/2015/02/03/383524379/new-york-state-clamps-down-on-herbal-supplements (accessed February 9, 2015).

19. Paul A. Offit, "Alternative Medicines Are Popular, But Do Any of Them Really Work?" *Washington Post*, November 11, 2013, http://www.washingtonpost.com/national/health-science/alternative-medicines-are-popular-but-do-any-of-them-really-work/2013/11/11/067f9272-004f-11e3-9711-3708310f6f4d_story.html (accessed February 12, 2015).

20. Center for Science in the Public Interest, "DNA Testing Reveals Herbal Supplement Often Missing the Advertised Supplement," February 3, 2015, http://cspinet.org/new/201502031.html (accessed February 7, 2015).

21. Anahad O'Connor, "Herbal Supplements Are Often Not What They Seem," *New York Times*, November 3, 2013, http://www.nytimes.com/2013/11/05/science/herbal-supplements-are-often-not-what-they-seem.html?pagewanted=all&_r=0 (accessed February 1, 2015).

22. Steven G. Newmaster, Meghan Grguric, Dhivya Shanmughanandhan, Sathishkumar Ramalingam, and Subramanyam Ragupathy, "DNA Barcoding Detects Contamination and Substitution in North American Herbal Products," *BMC Medicine* 11 (October 11, 2013): 222, doi:10.1186/1741-7015-11-222.

23. D. A. Baker, D. W. Stevenson, and D. P. Little, "DNA Barcode Identification of Black Cohosh Herbal Dietary Supplements," *Journal of AOAC International* 95, no. 4 (July/August 2012):102334.

24. Paul A. Offit and Sarah Erush, "Skip the Supplements," *New York Times*, December 14, 2013, http://www.nytimes.com/2013/12/15/opinion/sunday/skip-the-supplements.html (accessed February 7, 2015).

25. Ibid.

26. Ibid.

27. Christina Korownyk, Michael R. Kolber, James McCormack, Vanessa Lam, Kate Overbo, Candra Caitlin Finley, Ricky D. Turgeon, Scott Garrison, Adrienne J. Lindblad, Hoan Linh Banh, Denise Campbell-Scherer, Ben Vandermeer, and G. Michael Allan, "Televised Medical Talk Shows—What They Recommend and the Evidence to Support Their Recommendations: A Prospec-

tive Observational Study," *BMJ* (2014): 349, doi: http://dx.doi.org/10.1136/ bmj.g7346, http://www.bmj.com/content/349/bmj.g7346 (accessed January 19, 2015).

28. Elsevier, "Herbal Medicines Could Contain Dangerous Levels of Toxic Mold," *ScienceDaily*, October 23, 2014, www.sciencedaily.com/ releases/2014/10/141023091005.htm (accessed February 11, 2015).

29. Jim Lipton, "Support Is Mutual for Senator and Utah Industry," *New York Times*, June 20, 2011, http://www.nytimes.com/2011/06/21/us/ politics/21hatch.html?pagewanted=all&_r=0 (accessed February 10, 2015).

30. Ibid.

31. Ibid.

32. Pride Chigwedere, George R. Seage III, Sofia Gruskin, Tun-Hou Lee, and M. Essex, "Estimating the Lost Benefits of Antiretroviral Drug Use in South Africa," *Journal of Acquired Immune Deficiency Syndrome* 49, no. 4 (December 1, 2008): 410, http://www.aids.harvard.edu/Lost_Benefits.pdf (accessed December 22, 2014).

33. Paul A. Offit, "Alternative Healing or Quackery?" CNN, June 18, 2013, http://www.cnn.com/2013/06/18/health/alternative-medicine-offit/ (accessed February 5, 2015).

34. Steve Kroft, "Steve Jobs," CBS, October 20, 2011, www.cbsnews.com/ video/watch/?id=7385390n (accessed January 31, 2013).

CHAPTER 11: GOOD THINKING VS. BAD IDEAS

1. Nick Wynne, interview with the author, March 6, 2015.

2. Christopher Chabris and Daniel Simons, *The Invisible Gorilla and Other Ways Our Intuitions Deceive Us* (New York: Crown, 201), p. 160.

3. Michael Shermer, *The Believing Brain: From Ghosts and Gods to Politics and Conspiracies—How We Construct Beliefs and Reinforce Them as Truths* (New York: Times Books, 2011), p. 209.

4. Robin I. M. Dunbar, "How Conversations around Campfires Came to Be," *Proceedings of the National Academy of Sciences* 111, no. 39 (September 30, 2014): 14013–14.

5. University of Utah News Center, "Fire Talk of the Kalahari Bushmen," September 22, 2014, http://unews.utah.edu/news_releases/firelight-talk-of-the -kalahari-bushmen/ (accessed February 6, 2015).

6. Eric Anderson, Erika H. Siegel, Eliza Bliss-Moreau, and Lisa Feldman Barrett, "The Visual Impact of Gossip," *Science* 332, no. 6036, June 17, 2011, pp. 1446–48.

7. Ibid.

8. Ibid.

9. Robin Dunbar, *Grooming, Gossip and the Evolution of Language* (Cambridge, MA: Harvard University Press, 1996), pp. 4–5.

10. Public Policy Polling, "Democrats and Republicans Differ on Conspiracy Theory Beliefs," April 2, 2013, http://www.publicpolicypolling.com/pdf/2011/PPP_Release_National_ConspiracyTheories_040213.pdf (accessed February 7, 2015).

11. "Michele Bachmann Thinks Obama Is Causing the Apocalypse," *Huffington Post*, April 1, 2015, http://videos.huffingtonpost.com/michele-bachmann-thinks-obama-is-causing-the-apocalypse-518773558 (accessed June 6, 2015).

12. Centers for Disease Control, "Vaccines and Immunizations," http://www.cdc.gov/vaccines/vpd-vac/measles/ (accessed February 6, 2015).

13. Pew Research Center, "83% Say Measles Vaccine Is Safe for Healthy Children," February 9, 2015, http://www.people-press.org/2015/02/09/83-percent-say-measles-vaccine-is-safe-for-healthy-children/ (accessed May 2, 2015).

14. UNICEF, "Vaccines Bring 7 Diseases under Control," http://www.unicef.org/pon96/hevaccin.htm (accessed May 2, 2015).

15. Nick Triggle, "MMR Doctor Struck from Register," BBC News, May 24, 2010, http://news.bbc.co.uk/2/hi/health/8695267.stm (accessed February 7, 2015).

16. For a comprehensive list of such elected officials (and additional sources), see the *Wikipedia* page "Barack Obama Citizenship Conspiracy Theories," http://en.wikipedia.org/wiki/Barack_Obama_citizenship_conspiracy_theories (accessed June 15, 2015).

17. Caitlin Dickson, "Agenda 21: The U.N. Conspiracy That Just Won't Die," *Daily Beast*, April 13, 2014, http://www.thedailybeast.com/articles/2014/04/13/agenda-21-the-un-conspiracy-that-just-won-t-die.html (accessed June 6, 2015).

18. Ibid.

19. Senate Congressional Record, "Science of Climate Change," vol. 149, no. 113, July 28, 2003, pp. S10012–S10023, http://www.gpo.gov/fdsys/pkg/CREC-2003-07-28/html/CREC-2003-07-28-pt1-PgS10012.htm (accessed February 7, 2015).

20. Kurt Eichenwald, "The Plots to Destroy America," *Newsweek*, May 15, 2014, http://www.newsweek.com/2014/05/23/plots-destroy-america-251123.html (accessed February 3, 2015).

21. J. Eric Oliver and Thomas J. Wood, "Conspiracy Theories and the Paranoid Style(s) of Mass Opinion," *American Journal of Political Science* 58, no. 4 (October 2014): 952–66.

22. J. Eric Oliver and Tom Wood, "Larger Than Life," *New Scientist*, December 20–27, 2014, pp. 36–37.

23. Ibid.

24. Michael J. Wood, Karen M. Douglas, and Robbie M. Sutton, "Dead and Alive: Beliefs in Contradictory Conspiracy Theories," *Social Psychological and Personality Science* 3, no. 6767-773 (November 2012): 767–73.

25. Stephen E. Ambrose, "Writers on the Grassy Knoll," *New York Times Book Review*, February 2, 1992, http://www.nytimes.com/books/98/11/22/specials/ambrose-knoll.html (accessed February 26, 2015). Quoted in David Aaronovitch, *Voodoo Histories: The Role of the Conspiracy Theory in Shaping Modern History* (New York: Penguin Group, 2010), p. 356.

26. "Democrats and Republicans Differ on Conspiracy Theory Beliefs," Public Policy Polling, April 2, 2013, http://www.publicpolicypolling.com/pdf/2011/PPP_Release_National_ConspiracyTheories_040213.pdf (accessed February 7, 2015).

CHAPTER 12: OKAY, SO NOW WHAT?

1. Fred Gage, interview with the author, June 17, 2015.

2. Carey Goldberg, "Unlocking the Brain: Are We Entering a Golden Age of Neuroscience?" 90.9 WBUR: WBUR'S CommonHealth, http://commonhealth.wbur.org/2014/06/brain-matters-neuroscience-obama (accessed March 23, 2015).

3. John Pfister, interview with the author, February 22, 2015.

4. Carl Sagan and Ann Druyan, *Shadows of Forgotten Ancestors* (New York: Random House, 1992), p. 407.

5. Lane Kamp, interview with the author, April 27, 2015.

6. Graeme Wood, "What Isis Really Wants," *Atlantic*, March 2015, http://www.theatlantic.com/features/archive/2015/02/what-isis-really-wants/384980/ (accessed February 27, 2015).

7. Ibid.

8. BBC, "Joseph Kony: Profile of the LRA leader," March 8, 2012, http://www.bbc.com/news/world-africa-17299084 (accessed June 11, 2015).

BIBLIOGRAPHY

Aamodt, Sandra, and Sam Wang. *Welcome to Your Brain: Why You Lose Your Car Keys but Never Forget How to Drive, and Other Puzzles of Everyday Life*. New York: Bloomsbury, 2008.

Aaronovitch, David. *Voodoo Histories: The Role of the Conspiracy Theory in Shaping Modern History*. New York: Penguin Group, 2010.

Ackerman, Jennifer. *Sex Sleep Eat Drink Dream: A Day in the Life of Your Body*. New York: Houghton Mifflin Books, 2007.

Bader, Christopher, F. Carson Mencken, and Joseph Baker. *Paranormal America: Ghost Encounters, UFO Sightings, Bigfoot Hunts, and Other Curiosities in Religion and Culture*. New York: New York University Press, 2010.

Banaji, Mahzarin R., and Anthony G. Greenwald. *Blind Spot: Hidden Biases of Good People*. New York: Delacorte, 2013.

Barker, Dan. *Maybe Yes, Maybe No: A Guide for Young Skeptics*. Amherst, NY: Prometheus Books, 1990.

Barret, James. *The Future of the Brain: The Promise and Perils of Tomorrow's Neuroscience*. New York: Thomas Dunne Books, 2013.

Barrett, Stephen, and William T. Jarvis, eds. *The Health Robbers: A Close Look at Quackery in America*. Amherst, NY: Prometheus Books, 1993.

Bartholomew, Robert, and Benjamin Radford. *Hoaxes, Myths, and Manias: Why We Need Critical Thinking*. Amherst, NY: Prometheus Books, 2003.

Bausell, R. Barker. *Snake Oil Science: The Truth about Complementary and Alternative Medicine*. Oxford: Oxford University Press, 2009.

Bendheim, Paul E. *The Brain Training Revolution: A Proven Workout for Healthy Brain Aging*. Naperville, IL: Sourcebooks, 2009.

Blech, Jorg. *Healing through Exercise: Scientifically Proven Ways to Prevent and Overcome Illness and Lengthen Your Life*. Cambridge, MA: Da Capo, 2009.

Bloom, Floyd E., ed. *Best of the Brain from Scientific American: Mind, Matter, and Tomorrow's Brain*. New York: Dana, 2007.

Boghossian, Peter. *A Manual for Creating Atheists*. Charlottesville, VA: Pitchstone, 2013.

Bogin, Barry. *Patterns of Human Growth (Cambridge Studies in Biological and Evolutionary Anthropology)*. Cambridge, UK: Cambridge University Press, 1999.

Brafman, Ori, and Rom Brafman. *Sway: The Irresistible Pull of Irrational Behavior*. New York: Doubleday, 2008.

Brockman, John, ed. *The Mind: Leading Scientists Explore the Brain, Memory, Personality, and Happiness*. New York: Harper Perennial, 2011.

———. *Thinking: The New Science of Decision-Making, Problem-Solving, and Prediction*. New York: Harper Collins, 2013.

———. *This Will Make You Smarter: New Scientific Concepts to Improve Your Thinking*. New York: Harper Perennial, 2009.

———. *What Have You Changed Your Mind About?* New York: Harper Perennial, 2009.

Buonomano, Dean. *Brain Bugs: How the Brain's Flaws Shape Our Lives*. New York: W. W. Norton, 2011.

Burton, Frances. *Fire: The Spark That Ignited Human Evolution*. Albuquerque: University of New Mexico, 2009.

Burton, Robert A. *On Being Certain: Believing You Are Right Even When You're Not*. New York: St. Martin's, 2008.

———. *A Skeptic's Guide to the Mind: What Neuroscience Can and Cannot Tell Us about Ourselves*. New York: St. Martin's, 2013.

Calder, Nigel. *Magic Universe: A Grand Tour of Modern Science*. New York: Oxford University Press, 2003.

Capaldi, Nicholas, and Miles Smit. *The Art of Deception: An Introduction to Critical Thinking*. Amherst, NY: Prometheus Books, 2007.

Carey, Benedict. *How We Learn: The Surprising Truth about When, Where, and Why It Happens*. New York: Random House, 2014.

Carroll, Robert Todd, ed. *The Skeptic's Dictionary: A Collection of Strange Beliefs, Amusing Deceptions, and Dangerous Delusions*. Hoboken, NJ: John Wiley and Sons, 2003.

Carter, Rita. *The Human Brain Book*. New York: DK Adult, 2009.

Chabris, Christopher, and Daniel Simons. *The Invisible Gorilla and Other Ways Our Intuitions Deceive Us*. New York: Crown, 2010.

Chaffe, John. *Thinking Critically*. Boston: Houghton Mifflin, 2000.

Clancy, Susan A. *Abducted: How People Come to Believe They Were Kidnapped by Aliens*. Cambridge, MA: Harvard University Press, 2005.

Collen, Alanna. *10 % Human: How Your Body's Microbes Hold the Key to Health and Happiness*. New York: Harper, 2015.

Davis, Hank. *Caveman Logic: The Persistence of Primitive Thinking in a Modern World*. Amherst, NY: Prometheus Books, 2009.

Davis, James C. *The Human Story: Our History, from the Stone Age to Today*. New York: Harper Perennial, 2005.

Dawkins, Richard. *The Ancestor's Tale: A Pilgrimage to the Dawn of Evolution*. Boston: Houghton Mifflin, 2004.

———. *The Blind Watchmaker: Why the Evidence of Evolution Reveals a Universe without Design*. New York: W. W. Norton, 1996.

———. *Climbing Mount Improbable*. New York: W. W. Norton, 1997.

———. *The Greatest Show on Earth: The Evidence for Evolution*. New York: Free Press, 2009.

———. *The Magic of Reality: How We Know What's Really True.* New York: Free Press, 2011.

De Waal, Frans. *The Bonobo and the Atheist: In Search of Humanism among the Primates.* New York: W. W. Norton, 2013.

———. *Our Inner Ape: A Leading Primatologist Explains Why We Are Who We Are.* New York: Riverhead Books, Penguin Group, 2005.

Dennett, Daniel. *Consciousness Explained.* Boston: Little, Brown, 1991.

———. *Intuition Pumps and Other Tools for Thinking.* New York: W. W. Norton, 2014.

DeSalle, Rob, and Ian Tattersall. *The Brain: Big Bangs, Behaviors, and Beliefs.* New Haven and London: Yale University Press, 2012.

DiSalvo, David. *What Makes Your Brain Happy and Why You Should Do the Opposite.* Amherst, NY: Prometheus Books, 2011.

Dunbar, Robin. *Grooming, Gossip and the Evolution of Language.* Cambridge, MA: Harvard University Press, 1996.

Dunning, Brian. *Skeptoid: A Critical Analysis of Pop Phenomena.* Seattle, WA: Thunderwood, 2007.

———. *Skeptoid 2: More Critical Analysis of Pop Phenomena.* Seattle, WA: Skeptoid Media, 2008.

Eagleman, David. *Incognito: The Secret Lives of the Brain.* New York: Vintage, 2012.

Editors of *Scientific American.* *The Scientific American Book of the Brain.* New York: Lyon Press, 1999.

Feder, Kenneth L. *Encyclopedia of Dubious Archaeology: From Atlantis to the Walam Olum.* Santa Barbara, CA: Greenwood. 2010.

———. *Frauds, Myths, and Mysteries: Science and Pseudoscience in Archaeology.* Boston: McGraw-Hill, 2008.

Fernyhough, Charles. *Pieces of Light: How the New Science of Memory Illuminates the Stories We Tell about Our Pasts.* New York: Harper, 2013.

Feynman, Richard. *Surely You're Joking, Mr. Feynman!* New York: W. W. Norton, 1997.

Fine, Cordelia. *A Mind of Its Own: How Your Brain Distorts and Deceives.* New York: W. W. Norton, 2006.

Frazier, Kendrick. *Science under Siege: Defending Science, Exposing Pseudoscience.* Amherst, NY: Prometheus Books, 2009.

Freedman, Carl. *Conversations with Isaac Asimov.* Jackson: University Press of Mississippi, 2005.

Garreau, Joel. *Radical Evolution: The Promise and Peril of Enhancing Our Minds, Our Bodies— and What It Means to Be Human.* New York: Doubleday, 2005.

Goldacre, Ben. *Bad Science: Quacks, Hacks, and Big Pharma Flacks.* New York: Faber and Faber, 2010.

Gottshcall, Jonathan. *The Storytelling Animal: How Stories Make Us Human.* New York: Houghton Mifflin Harcourt, 2012.

Gould, Stephen Jay. *The Flamingo's Smile: Reflections in Natural History*. New York: W. W. Norton, 1985.

———. *Ever Since Darwin: Reflections in Natural History*. New York: W. W. Norton, 1992.

———. *Hen's Teeth and Horse's Toes: Further Reflections in Natural History*. New York: W. W. Norton, 1983.

———. *Wonderful Life: The Burgess Shale and the Nature of History*. New York: W. W. Norton, 2007.

Hanlon, Michael. *Eternity: Our Next One Billion Years*. London: Macmillan, 2009.

Harrison, Guy P. *50 Popular Beliefs That People Think Are True*. Amherst, NY: Prometheus Books, 2012.

———. *50 Reasons People Give for Believing in a God*. Amherst, NY: Prometheus Books, 2008.

———. *50 Simple Questions for Every Christian*. Amherst, NY: Prometheus Books, 2013.

———. *Race and Reality: What Everyone Should Know about Our Biological Diversity*. Amherst, NY: Prometheus Books, 2010.

———. *Think: Why You Should Question Everything*. Amherst, NY: Prometheus Books, 2013.

Hart, Donna. *Man the Hunted: Primates, Predators, and Human Evolution*. New York: Basic Books, 2005.

Haught, James A. *Holy Hatred: Religious Conflicts of the '90s*. Amherst, NY: Prometheus Books, 1995.

———. *Honest Doubt: Essays on Atheism in a Believing Society*. Amherst, NY: Prometheus Books, 2007.

———. *2,000 Years of Disbelief: Famous People with the Courage to Doubt*. Amherst, NY: Prometheus Books, 1996.

Head, Tom, ed. *Conversations with Carl Sagan*. Jackson: University Press of Mississippi, 2006.

Hedges, Chris. *Empire of Illusion: The End of Literacy and the Triumph of Spectacle*. New York: Nation Books, 2010.

Hemenway, Priya. *Hindu Gods: The Spirit of the Divine*. San Francisco: Chronicle Books, 2003.

Hines, Terrence. *Pseudoscience and the Paranormal*. Amherst, NY: Prometheus Books, 2003.

Horstman, Judith. *The Scientific American Brave New Brain*. San Francisco: Jossey-Bass, 2010.

———. *The Scientific American Day in the Life of Your Brain*. San Francisco: Jossey-Bass, 2009.

Huffman, Karen. *Psychology in Action*. Hoboken, New Jersey: John Wiley and Sons, 2007.

Humes, Edward. *Monkey Girl: Evolution, Education, Religion, and the Battle for America's Soul*. New York: HarperCollins, 2007.

Jarrett, Christian. *Great Myths of the Brain*. West Sussex, UK: John Wiley and Sons, 2015.

Jordan, Michael. *Encyclopedia of Gods*. London: Kyle Cathie, 2002.

Kaku, Michio. *The Future of the Mind: The Scientific Quest to Understand, Enhance, and Empower the Mind*. New York: Doubleday, 2014.

Kelly, Lynne. *The Skeptic's Guide to the Paranormal*. New York: Avalon, 2004.

Kida, Thomas. *Don't Believe Everything You Think: The 6 Basic Mistakes We Make in Thinking*. Amherst, NY: Prometheus Books, 2006.

Kurtz, Paul. *Affirmations: Joyful and Creative Exuberance*. Amherst, NY: Prometheus Books, 2004.

———. *The New Skepticism: Inquiry and Reliable Knowledge*. Amherst, NY: Prometheus Books, 1992.

———. ed. *Science and Religion: Are They Compatible?* Amherst, NY: Prometheus Books, 2003.

———. ed. *Skeptical Odysseys: Personal Accounts by the World's Leading Paranormal Inquirers*. Amherst, NY: Prometheus Books, 2001.

———. *The Transcendental Temptation: A Critique of Religion and the Paranormal*. Amherst, NY: Prometheus Books, 2013.

Kurzweil, Ray. *How to Create a Mind: The Secret of Human Thought Revealed*. New York: Penguin, 2012.

———. *The Singularity Is Near: When Humans Transcend Biology*. New York: Penguin, 2006.

Leeming, David. *A Dictionary of Creation Myths*. New York: Oxford University Press, 1994.

Levitin, Daniel. *This Is Your Brain on Music: The Science of a Human Obsession*. New York: Penguin, 2006.

———. *The World in Six Songs: How the Musical Brain Created Human Nature*. New York: Plume, 2008.

Lieberman, Daniel E. *The Evolution of the Human Head*. Cambridge, MA: Belknap Press of Harvard University Press, 2011.

———. *The Story of the Human Body: Evolution, Health, and Disease*. New York: Pantheon Books, 2013.

Loftus, John W. *The End of Christianity*. Amherst, NY: Prometheus Books, 2013.

———. *The Outsider Test for Faith: How to Know Which Religion Is True*. Amherst, NY: Prometheus Books, 2013.

Loxton, Daniel. *Evolution: How We and All Living Things Came to Be*. Tonawanda, NY: Kids Can Press, 2010.

Lynch, John, and Louise Barrett. *Walking with Cavemen*. New York: DK Publishing, 2003.

Mackay, John. *Extraordinary Popular Delusions and the Madness of Crowds*. Hertfordshire, England: Wordsworth Editions, 1999.

Macknic, Stephen L., and Susana Martinez-Conde. *Sleights of Mind: What the*

Neuroscience of Magic Reveals about Our Everyday Deceptions. New York: Henry Holt, 2010.

Marcus, Gary. *Kluge: The Haphazard Construction of the Human Mind*. New York: Houghton Mifflin, 2008.

Mayr, Ernst. *What Evolution Is*. London: Weidenfeld and Nicolson, 2002.

McLaren, Carrie, and Jason Torchinsky. *Ad Nauseam: A Survivor's Guide to American Consumer Culture*. New York: Faber and Faber, 2009.

McRaney, David. *You Are Now Less Dumb: How to Conquer Mob Mentality, How to Buy Happiness, and All the Other Ways to Outsmart Yourself*. New York: Gotham. 2013.

———. *You Are Not So Smart: Why You Have Too Many Friends on Facebook, Why Your Memory Is Mostly Fiction, and 46 Other Ways You're Deluding Yourself*. New York: Gotham, 2012.

Medina, John. *Brain Rules: 12 Principles for Surviving and Thriving at Work, Home, and School*. Seattle: Pear, 2008.

Mills, David. *Atheist Universe: The Thinking Person's Answer to Christian Fundamentalism*. Berkeley, CA: Ulysses, 2006.

Mnookin, Seth. *The Panic Virus: A True Story of Medicine, Science, and Fear*. New York: Simon and Schuster, 2011.

Mooney, Chris, and Sheril Kirshenbaum. *Unscientific America: How Scientific Illiteracy Threatens Our Future*. New York: Basic Books, 2009.

Murdoch, Stephen. *IQ: A Smart History of a Failed Idea*. Hoboken, NJ: Wiley and Sons, 2007.

Musolino, Julien. *The Soul Fallacy: What Science Shows We Gain from Letting Go of Our Soul Beliefs*. Amherst, NY: Prometheus Books, 2015.

Naam, Ramez. *More Than Human: Embracing the Promise of Biological Enhancement*. New York: Broadway Books, 2005.

National Academy of Sciences. *Science, Evolution, and Creationism*. Washington, DC: National Academies, 2008.

Nelson, Kevin. *The Spiritual Doorway in the Brain: A Neurologist's Search for the God Experience*. New York: Dutton, 2011.

Nickell, Joe. *Adventures in Paranormal Investigation*. Lexington: University Press of Kentucky, 2007.

———. *Looking for a Miracle: Weeping Icons, Relics, Stigmata, Visions & Healing Cures*. Amherst, NY: Prometheus Books, 1999.

———. *The Mystery Chronicles: More Real-Life X-Files*. Lexington: University Press of Kentucky, 2004.

———. *Psychic Sleuths: ESP and Sensational Cases*. Amherst, NY: Prometheus Books, 1994.

———. *Relics of the Christ*. Lexington: University Press of Kentucky, 2007.

Nisbett, Richard E. *Intelligence and How to Get It: Why Schools and Cultures Count*. New York: W. W. Norton, 2009.

Offit, Paul A. *Autism's False Prophets: Bad Science, Risky Medicine, and the Search for a Cure*. New York: Columbia University Press, 2010.

———. *Bad Faith: When Religious Belief Undermines Modern Medicine*. New York: Basic Books, 2015.

———. *Deadly Choices: How the Anti-Vaccine Movement Threatens Us All*. New York: Basic Books, 2010.

———. *Do You Believe in Magic? Vitamins, Supplements, and All Things Natural: A Look behind the Curtain*. New York: Harper Paperbacks, 2014.

Offit, Paul A., and Charlotte A. Moser. *Vaccines and Your Child: Separating Fact from Fiction*. New York: Columbia University Press, 2011.

Palmer, Douglas. *Origins: Human Evolution Revealed*. New York: Mitchell Beazley, 2010.

Park, Robert. *Voodoo Science: The Road from Fraud to Foolishness*. New York: Oxford University Press, 2000.

Pigliucci, Massimo. *Nonsense on Stilts: How to Tell Science from Bunk*. Chicago: University of Chicago Press, 2010.

Piper, Don. *90 Minutes in Heaven: A True Story of Death and Life*. Grand Rapids, MI: Revell, 2004.

Plait, Philip. *Bad Astronomy: Misconceptions and Misuses Revealed, from Astrology to the Moon Landing "Hoax."* New York: John Wiley and Sons, 2002.

Prothero, Stephen. *Religious Literacy: What Every American Needs to Know— And Doesn't*. New York: HarperOne, 2007.

Radford, Benjamin. *Media Mythmakers: How Journalists, Activists and Advertisers Mislead Us*. Amherst, NY: Prometheus Books, 2003.

———. *Scientific Paranormal Investigation: How to Solve the Unexplained Mysteries*. Corrales, NM: Rhombus, 2010.

———. *Tracking the Chupacabra: The Vampire Beast in Fact, Fiction, and Folklore*. Albuquerque: University of New Mexico Press, 2011.

Randi, James. *An Encyclopedia of Claims, Frauds, and Hoaxes of the Occult and Supernatural*. New York: St. Martin's Griffin, 1995.

———. *The Faith Healers*. Amherst, NY: Prometheus Books, 1989.

———. *Flim-Flam! Psychics, ESP, Unicorns, and Other Delusions*. Amherst, NY: Prometheus Books, 1982.

———. *The Mask of Nostradamus: The Prophecies of the World's Most Famous Seer*. Amherst, NY: Prometheus Books, 1993.

Ramachandran, V. S. *The Tell-Tale Brain: A Neuroscientist's Quest for What Makes Us Human*. New York: W. W. Norton, 2011.

Ratey, John J. *Spark: The Revolutionary New Science of Exercise and the Brain*. New York: Little, Brown, 2013.

———. *A User's Guide to the Brain: Perception, Attention, and the Four Theaters of the Brain*. New York: Vintage, 2002.

Reynolds, Gretchen. *The First 20 Minutes: Surprising Science Reveals How We Can Exercise Better, Train Smarter, Live Longer*. New York: Hudson Street, 2012.

Rose, Steven. *The Future of the Brain: Essays by the World's Leading Neuroscientists*. New York: Oxford University Press, 2005.

Sacks, Oliver. *Hallucinations*. New York: Alfred A. Knopf/Random House, 2013.

———. *Musicophilia: Tales of Music and the Brain*. New York: Vintage, 2008.

Sagan, Carl. *Billions & Billions: Thoughts on Life and Death at the Brink of the Millennium*. New York: Ballantine, 1998.

———. *The Demon-Haunted World: Science as a Candle in the Dark*. New York: Random House, 1995.

———. *Pale Blue Dot: A Vision of the Human Future in Space*. New York: Random House, 1994.

———. *The Varieties of Scientific Experience: A Personal View of the Search for God*. New York: Penguin, 2007.

Sagan, Carl, and Ann Druyan. *Shadows of Forgotten Ancestors: A Search for Who We Are*. New York: Random House, 1992.

Satel, Sally, and Scott O. Lilienfeld. *Brainwashed: The Seductive Appeal of Mindless Neuroscience*. New York: Basic Books, 2013.

Sawyer, G. J., and Victor Deak. *The Last Human: A Guide to Twenty-Two Species of Extinct Humans*. New Haven: Yale University Press, 2007.

Schick, Theodore, and Lewis Vaughn. *How to Think about Weird Things*. New York: McGraw-Hill, 2011.

Scott, Eugenie C. *Evolution vs. Creationism: An Introduction*. Berkeley: University of California Press, 2009.

Seung, Sebastian. *Connectome: How the Brain's Wiring Makes Us What We Are*. New York: Houghton Mifflin Harcourt, 2012.

Shapiro, Rose. *Suckers: How Alternative Medicine Makes Fools of Us All*. London: Harvill Secker, 2008.

Sheaffer, Robert. *UFO Sightings: The Evidence*. Amherst, NY: Prometheus Books, 1998.

Shenk, David. *The Genius in All of Us: New Insights into Genetics, Talent, and IQ*. New York: Doubleday, 2010.

Shermer, Michael. *The Believing Brain: From Ghosts and Gods to Politics and Conspiracies—How We Construct Beliefs and Reinforce Them as Truths*. New York: Times Books, 2011.

———. *The Borderlands of Science: Where Sense Meets Nonsense*. New York: Oxford University Press, 2002.

———. *Science Friction: Where the Known Meets the Unknown*. New York: Times Books, 2005.

———. *Why Darwin Matters: The Case against Intelligent Design*. New York: Times Books, 2006.

———. *Why People Believe Weird Things: Pseudoscience, Superstition, and Other Confusions of Our Time*. New York: MJF Books, 1997.

Shostak, Seth. *Confessions of an Alien Hunter: A Scientist's Search for Extraterrestrial Intelligence*. Washington, DC: National Geographic, 2009.

Shubin, Neil. *Your Inner Fish: A Journey into the 3.5-Billion-Year History of the Human Body*. New York: Pantheon Books, 2008.

Singh, Simon, and Edzard Ernst. *Trick or Treatment: The Undeniable Facts about Alternative Medicine*. New York: W. W. Norton, 2008.

Smith, Cameron M. *The Fact of Evolution*. Amherst, NY: Prometheus Books, 2011.

Smith, Cameron M., and Charles Sullivan. *The Top 10 Myths about Evolution*. Amherst, NY: Prometheus Books, 2006.

Smith, Jonathan C. *Pseudoscience and Extraordinary Claims of the Paranormal: A Critical Thinker's Toolkit*. West Sussex, UK: Wiley-Blackwell, 2010.

Specter, Michael. *Denialism: How Irrational Thinking Hinders Scientific Progress, Harms the Planet, and Threatens Our Lives*. New York: Penguin, 2009.

Stanovich, Keith E. *How to Think Straight about Psychology*. New York: HarperCollins, 1996.

Stenger, Victor J. *The Fallacy of Fine-Tuning: Why the Universe Is Not Designed for Us*. Amherst, NY: Prometheus Books, 2011.

———. *God and the Multiverse: Humanity's Expanding View of the Cosmos*. Amherst, NY: Prometheus Books, 2014.

———. *The New Atheism: Taking a Stand for Science and Reason*. Amherst, NY: Prometheus Books, 2009.

Stringer, Chris. *Lone Survivors: How We Came to Be the Only Humans on Earth*. New York: Times Books, 2011.

Stringer, Chris, and Peter Andrews. *The Complete World of Human Evolution*. New York: Thames and Hudson, 2005.

Suzuki, Wendy, with Billie Fitzpatrick. *Healthy Brain, Happy Life: A Personal Program to Activate Your Brain and Do Everything Better*. New York: Harper Collins, 2015.

Sweeney, Michael S. *Brain: The Complete Mind: How It Develops, How It Works, and How to Keep It Sharp*. Washington, DC: National Geographic, 2009.

———. *Brain Works: The Mind-Bending Science of How You See, What You Think, and Who You Are*. Washington, DC: National Geographic, 2011.

Tattersall, Ian. *Becoming Human: Evolution and Human Uniqueness*. New York: Harcourt Brace, 1998.

———. *Extinct Humans*. New York: Basic Books, 2001.

———. *The Human Odyssey: Four Million Years of Human Evolution*. New York: Prentice Hall, 1993.

———. *Masters of the Planet: The Search for Our Human Origins*. New York: Palgrave Macmillan, 2012.

Thompson, Damian. *Counterknowledge: How We Surrendered to Conspiracy Theories, Quack Medicine, Bogus Science, and Fake History*. New York: W. W. Norton, 2008.

Tyson, Neil DeGrasse, and Donald Goldsmith. *Origins: Fourteen Billion Years of Cosmic Evolution.* New York: W. W. Norton, 2004.

Van Hecke, Madeleine. *Blind Spots: Why Smart People Do Dumb Things.* Amherst, NY: Prometheus Books, 1997.

Wanjek, Christopher. *Bad Medicine: Misconceptions and Misuses Revealed, from Distance Healing to Vitamin O.* New York: Wiley, 2002.

Weinberg, Steven. *Facing Up: Science and Its Cultural Adversaries.* Cambridge, MA: Harvard University Press, 2003.

———. *Lake Views: This World and the Universe.* Cambridge, MA: Belknap Press of Harvard University Press, 2010.

Wilson, Edward O. *The Meaning of Existence.* New York: W. W. Norton, 2014.

———. *The Social Conquest of Earth.* New York: Liveright Publishing, 2012.

Wiseman, Richard. *Paranormality: Why We See What Isn't There.* London: Macmillan, 2011.

Woerlee, G. M. *Mortal Minds: The Biology of Near-Death Experiences.* Amherst, NY: Prometheus Books, 2005.

Wrangham, Richard. *Catching Fire: How Cooking Made Us Human.* New York: Basic Books, 2009.

Young, Matt, and Taner Edis. *Why Intelligent Design Fails: A Scientific Critique of the New Creationism.* Piscataway, NJ: Rutgers University Press, 2006.

Zimmer, Carl. *Evolution: The Triumph of an Idea.* New York: Harper Perennial, 2006.

———. *Soul Made Flesh: The Discovery of the Brain—and How It Changed the World.* New York: Free Press, 2004.

Zott, Lynn M. *Alternative Medicine (Opposing Viewpoints).* Farmington Hills, MI: Greenhaven, 2012.

Zuckerman, Phil. *Faith No More: Why People Reject Religion.* New York: Oxford University Press, 2011.

———. *Society without God: What the Least Religious Nations Can Tell Us about Contentment.* New York: NYU Press, 2010.

ABOUT THE AUTHOR

Guy P. Harrison is a passionate advocate for science and reason who says he hates to see people suffer unnecessarily. He calls the challenge of good thinking a moral issue and points to poor reasoning as humankind's great unrecognized crisis. Guy enjoys sharing his positive, constructive style of skepticism and science appreciation with people whenever possible. He has a degree in history and anthropology and has visited more than twenty countries on six continents. Having seen some of the best and worst of our world, he believes that we can do better. Guy maintains that if more people embraced critical thinking and had a better understanding of basic brain processes such as sensory perception, memory, and subconscious biases, we could eliminate a significant amount of human suffering and become much more efficient, safer, and productive as a species.

As a journalist, Guy has worked in many roles, including editorial writer, world-news editor, sports editor, reporter, feature writer, and columnist. He won the Commonwealth Award for Excellence in Journalism and the WHO (World Health Organization) Award for Health Reporting. Guy has also interviewed many leading scientists and significant historical figures. He has written about many diverse topics, including poverty in the developing world, conservation issues, religion, war, racism, gender discrimination, space exploration, and human origins.

Although he says he's an introvert, Guy never misses a chance to spread science and reason with others. He has been a guest on more than one hundred radio shows and podcasts and was a featured speaker at a science festival in New Zealand and at a Random House conference in San Diego, California.

Guy is the author of five previous books that have been popular with readers and highly acclaimed by critics. They are: *Think: Why You Should Question Everything*; *50 Simple Questions for*

Every Christian; *50 Popular Beliefs That People Think Are True*; *50 Reasons People Give for Believing in a God;* and *Race and Reality: What Everyone Should Know about Our Biological Diversity*. Random House selected *Think* as part of its national First Year Experience/Common Reads program, which promotes it as recommended reading for first-year university students.

Guy is a lifelong fan not only of science and history but also of science fiction. He says he's not ashamed to confess his deep love for robot uprisings, time machines, and interstellar travel. He lives in Southern California, where he enjoys running, hiking, biking, and writing.

Guy is also an "expert blogger" for *Psychology Today*. Read his essays at *About Thinking*, www.psychologytoday.com/blog/about -thinking. Visit his website at www.guypharrison.com.

INDEX

Page numbers set in **boldface** indicate images.